Direct to Brain Windows

Remote Neuron Reading and Writing and Other Science
Big Secrets, Lies, and Mistakes

Ted Huntington

Direct to Brain Windows, Remote Neuron Reading and Writing and Other Science Big Secrets, Lies, and Mistakes
Copyright © 2012 by Ted Huntington

ISBN (978-0-9881922-0-1)

www.tedhuntington.org

Dedicated to the many people excluded from direct-to-brain windows and victims of remote neuron writing everywhere (if such things exist)

CONTENTS

Chapter 1
Introduction

If you are reading this book, than you are probably "excluded", that is, you don't get direct-to-brain windows, or direct-to-eyes videos, and you have to read books – you don't see semi-transparent windows in front of your eyes right now- like the people on the cover of this book.

I understand that this sounds like crazy talk to most people who have never heard about remote neuron reading and writing, but as you will see, there is a lot of evidence to support the idea that some humans invented, developed, and perfected remote neuron reading and writing a long time ago, but chose not to tell the public. Even if you doubt this, you have to admit that receiving videos directly to our eyes and ears would be much more convenient than having to look at a screen or use headphones, and that seeing and hearing the images and sounds of thought would relieve us from having to repeat our thought-audio out loud and to reproduce our thought-images graphically on paper. I myself, believe with a strong amount of certainty, from the evidence I will present, and from my own personal experience, that many people are being provided with some kind of direct-to-brain windows, while many others, including myself, are being mostly denied this basic service, as unbelievable as that may sound.

I feel tremendously for the "excluded" and for all my fellow humans. I can't believe what has happened, and what is being done to people, many of whom are very gentle and decent people, and even many other species, equally gentle and thought-filled. This is why I have dedicated many hours to trying to show the public the truth about direct-to-brain windows and remote neuron reading and writing.

I am excluded and have been my entire life of 43 years. Many millions of people are being denied; my mother, for example, was born, lived, and died never receiving direct to brain windows, other than the usual millisecond tiny images or sounds that most excluded dismiss or think are internally created.

I have an unbelievable advantage that most other people do not have, and that is that I know a lot about remote neuron reading and writing ("RNRAW") and direct-to-brain windows ("D2BW" or simply "D2B"), but yet, I am still excluded. I don't receive direct-to-brain windows. Many, and perhaps even most people around me do receive D2B, but of course, all D2B consumers are held as hostages by the direct-to-brain owners and controllers, and mostly remain silent about whatever videos they are seeing in their eyes. So I am one of the few people that can actually tell the public a little about remote neuron reading and writing. Those who are excluded have never thought about it primarily because the major media (newspapers, television, etc.) and schools, the source of most of what we know, are not telling the public about D2B. Those who are included can't talk publicly about it for fear of losing their D2B service, and because of the tremendous advantages (jobs, mates, safety, etc.) that seeing and hearing inside people's houses and heads gives to most people. I know a lot about D2B, but I don't receive D2B, so I can talk pretty much openly about D2B, because I can't lose a service I don't receive.

If you do not receive direct-to-brain windows, then most of the amazing things I am about to tell you in this book, you probably have never heard of before. So it is no surprise that some of the claims, like the claim that tiny floating and flying microscopic cameras fill the air of Earth may, at first, seem crazy, or seem possible, but only far in the future. But it seems likely to me that those who get direct-to-brain windows have known all these simple truths for centuries. After you learn about D2B, many of the mistakes of your past, and what other people say and do around you make much more sense. Beyond that, there is absolutely no harm in thinking about such things.

One of the biggest problems with reaching people who do not receive direct-to-brain windows and know nothing about remote neuron reading and writing is that their eyes "glaze over" when hearing talk about mind reading and telepathy. I know from experience, that even my own eyes glaze over. People quickly dismiss any such talk as pseudoscience and mysticism- and most of it is. It's important to distinguish that D2B windows and RNRAW is done through science, is done using technology, and may have been kept secret for many years. Another problem is that people always say "hearing thought- oh that's hundreds of years away in the future" and the evidence suggests that the technology of hearing thought *is* hundreds of years away...but hundreds of years away *in the past*. So it's important to say up front that this is not about the mystical traditional "telepathy" association- although, as the timeline I give you later shows, many of those people associated with "telepathy" (like William Crookes and Upton Sinclair) clearly were hinting about this massive secret technology and shocking segregation.

What we are seeing, and will be seeing for many years to come, is what I call "The Big Lie", a massive group of humans that form a big all-powerful political party, that selfishly hog D2BW to themselves, live as D2B addicts with a constant D2B "fear" paralysis, and constantly echo these same old childish, obviously false party-line lies and theories- like "19-hijackers did 9/11", "Oswald killed JFK", "Jesus rose from the dead", "the universe is expanding", "people don't hear your thoughts", etc. But, it seems inevitable that the truth will ultimately, through many tortuous centuries, finally emerge, and win over the minds of Earth.

Chapter 2
Direct to Brain Windows Basics

Explaining what "Direct to Brain Windows", or more formally "Remote Neuron Reading and Writing", is, and how it works takes some time because there are a lot of details. It is initially difficult to understand and visualize for some people who have never received direct-to-brain windows.

Wouldn't it be nice to be able to walk around and have the Internet right there in front of our eyes without needing to look at a screen? We could look at our calendar, look at the time and weather forecast, right there on the spot, as semi-transparent windows in front of our eyes. It is similar to if you have ever looked at bright sunlight through window blinds and then turned away to still see the image of the blinds in your eyes. Imagine that in the same way as still seeing those window shades after you turned away from the window, you could see semi-transparent videos and windows. You could watch a movie while you exercise for example. These windows could even have the Internet, and include the ability to search by drawing words you want to search for in the window with your mind. All the people around you would have no idea that you were watching movies and searching the Internet, because all they would see is you as you usually look- without any semi-transparent windows in front of you.

But then how about seeing not just the Internet in front of our eyes wherever we look and everywhere we go, but an Internet that extremely intrudes on the idea of privacy – playing you videos from inside people's rooms, that even plays you the sounds of their thought, and shows you the very images they are thinking of? And an Internet that shows you videos of their past: their most violent crimes, their most embarrassing moments, their most sexual events, all kinds of interesting and important information about them. It would be very informative, interesting and helpful. But then imagine that one group of people gets this service, and another group of people is denied it.

In this book I will be describing much of what I have learned about remote neuron reading and writing, and also many basic science truths that, like direct-to-brain windows, have been kept secret from the public.

Remote neuron reading and writing has not yet been acknowledged publicly, but we should understand the technology in preparation for a time when it is made public, and also in the interest of our own safety, and for our own knowledge, if remote neuron reading and writing is already in use secretly.

What is the nature of direct-to-brain windows? How might it work?

It may seem complex and overwhelming, but it is somewhat simple. The universe is much more easy to understand when learning two simple facts: All matter is made of material light particles, and matter and motion can never be created, exchanged or destroyed. But even this simple truth has been kept from the public.

One funny aspect is how some evidence of remote neuron reading and writing has been in plain sight, but many of us excluded simply don't notice it or think about it. For example, there are many times in movies and television shows where you hear one of the characters thoughts. In cartoons, bubbles leading up to an Image of what the character is thinking are common. It's interesting to realize that, those comical images of cartoon characters with a thought screen turn out to be a major secret paradigm- that, in fact, from the perspective of many D2B consumers, we all now walk around with those bubbles and our thought screen (and a screen showing what our eyes are looking at) permanently, to be viewed all the time! Here we saw it in cartoons for a century and most people didn't make the connection- I certainly never did until learning about D2B. And then, one of the sneakiest parts is that the

D2B consumers don't even need to look over your head to see your thought-screen, they can get it sent directly to their brain (and eyes) without even being near you.

The basis for D2B windows and remote neuron reading and writing, are basic devices: many, many **tiny particle devices**. These devices can function as **cameras**, **microphones**, and particle **communication devices**. For a poor person who must work much of their life, it is difficult to understand the perspective from the eyes of a person who doesn't have to work, because they are wealthy enough to do whatever they want to do all the time. Very wealthy people have much more time and money to devote to developing technology than most of us do. In addition, most of their friends are very wealthy and have the same benefits they do, so they fit in with a different society- they don't really mix with the majority of average middle income people. So over the centuries you can see how technology can be improved and improved, secretly, so that it may be light-years from what the public is familiar with. However, over the centuries, somehow, mysteriously, more and more poor and middle income people have been permitted to receive direct-to-brain windows- but most of the public remains completely unaware of it.

Experience probably showed the wealthy leaders of many nations very early on that seeing and hearing what the leaders of other nations are doing is extremely useful in defending yourself, and preventing any surprise attacks. Even locally, seeing a live image and hearing the messages of your enemies and even your friends, in particular when they do not see you, can give a person a tremendous advantage in survival- for example to know if there are plans of violence and just simply to know what the truth of many situations are. In addition, many new and interesting science facts may be obtained from seeing what other people do. But, no doubt one driving force of filling the planet with tiny dust sized cameras, microphones and particle communication devices is the driving desire to see pretty women by males.

Initially, the first cameras and microphones must have been very crude and large. The public understanding of cameras really only starts with Niépce and the first public photograph in 1816, but it seems very likely, like the Manhattan Project, and given many of the examples that I present later in this book, that particle devices for communication, as weapons, and for large-scale planetary observation first occurred and were developed probably much, much earlier than the public has been told. There is absolutely no harm in exploring this possibility. If true, this secret development probably quickly resulted in a desire for smaller and smaller communication, camera and microphone devices. It may surprise people to realize that small sand-grain sized devices do not have to be very small to go unnoticed by the average human. Simply look under a bright desk lamp at all the tiny dust fibers floating around; they are always there, but we don't normally notice them. How far away from your eyes does a grain of salt (perhaps 1mm in size) need to be before you can't see it anymore? Not very far. All that is really needed is a device that is perhaps a micrometer in size that is powered by ambient light particles, which can capture images and sounds and send them to other devices. This tiny dust-sized device could float in the air of earth, or even have tiny motors (microscopic motors have been public since at least 1988- see the timeline) to fly around.

Beyond this, wealthy people must have learned early on what a "neuron" is and how they work. Neurons are the cells that form all nerves – all animals have them. They are like transistors in being electrical switches that can be viewed as either "on" or "off". Like all matter, neurons absorb and emit light particles. So like an electronic memory, a neuron can be read from and written to. In a very real sense, the brain of any species is a lot like a computer memory.

"People don't watch us in our homes!"

Like many famous theories and claims of history, this belief has apparently been very mistaken, and much of the public's trust very misplaced. If you think that nobody is capturing videos of you no matter where you are on the tiny planet Earth, this may be a very serious mistake, because you may act out on some suggestion to do some violent or nonviolent crime thinking that nobody can see you.

"I don't want to see people in their homes!"

This may be a common reaction too, but once you see how many other people already are seeing, then you realize that this is just the natural advance of modern communication technology. You realize, that the nude human body is not as shocking and embarrassing as many people have thought, and that the benefits of getting important information far outweigh the loss of privacy. The ancient concept of "privacy", as shocking as it may seem, is actually a myth now, because of the advent of remote neuron reading and writing and nanotechnology. The only people that might enjoy privacy are those who own and control the remote neuron technology- those that watch millions of people and their thoughts all day and night- and those are the last people that should have any privacy.

"I don't want people to see and hear my thoughts!"

For many people this may be the initial reaction. There is a feeling that somehow we are embarrassed about our thoughts. But when everybody can see each other's thought screen and hear each other's thought-audio, you can see how the view is that this is just one of those advances in technology that makes our lives much easier, but also introduces some changes- just like the spinning-Jenny, the automobile, atomic fission, the x-ray, and the telephone did. We need to adapt to this new reality, and the benefits far outweigh any unpleasantness. We can't go back to the stone age, and can only move forward into the exciting inevitable future.

"Shouldn't some people be excluded from seeing inside houses and heads?"

One of the big problems with how remote neuron reading and writing evolved on Earth, is that it somehow fell into the hands of people who are somewhat violent and don't care a lot about science, education, or other people. As a result, many millions of fine nonviolent people are not only excluded from being able to see, hear and send images and sounds to and from brains, but have been told absolutely nothing about remote neuron reading and writing.

Those people who do get D2B have an unbelievable advantage over those who are "excluded" from D2B in being able to rapidly communicate with other D2B consumers, and to see and hear thoughts instantly. Beyond that, those who are excluded from D2B are the victim of terrible suggestions and molestations that are written invisibly, remotely, and unconsensually to their brains. Their muscles are remotely moved, they are made to itch, to hear unpleasant put-downs and suggestions in their thought-audio, etc. The vast majority don't realize that some unseen violent criminals are sending sounds or images remotely to their brain – many of them interpret the sounds or images as messages from God, and promptly do what the voice or image suggests to them even if it is a suggestion to lie, steal, do something sexually inappropriate, or to do violence. The view may be that they obviously do not want to disobey the voice of, or message from God, no matter how terrible the request God is making may be.

I think that some people should be excluded from receiving D2B- but only the very violent. In particular I vote to exclude people with involvement in multiple or serious acts of first strike violence against non-violent people. Certainly, those who repeatedly remotely write violent or unwanted neuron writings should lose the "right to write" to at least the neurons of their victims and most likely any and all other humans. But sadly, what we see is a very different "pecking order".

You need to understand the historical context in which we live, and in which remote neuron reading and writing were born. Throughout much of history, the majority of humans have been ruled by inaccurate superstitious beliefs which were violently enforced on the society. The persecutions of Anaxagoras and Socrates for disrespecting the Gods are clear pre-Christian examples of the dominance of traditional superstitious theories over more honest and accurate alternative theories. The rise of Christianity around the 3rd and 4th centuries AD set back the development of science through the Dark Ages and this continued through the years of the Inquisition and witch trials into modern times. Only in the last few centuries have scientific truths such as the Heliocentric theory and the theory of Evolution started to become publicly known and those teaching these theories free from punishment and violence. So it is perhaps no wonder that much of remote neuron reading and writing has fallen into the same traditional, superstitious, and violent hands that have dominated the planet Earth for centuries. Right now, generally speaking, as has been the case for most of the history of Earth, good is at the bottom and evil is on the top, victims on the bottom, murderers on the top. In the current tradition, the "Galileos" (and "Galileas") of Earth, those who care deeply for truth, see other people as worthy of knowing the truth, and want to teach the public scientific truths are most often excluded, poor and powerless, while people who lie to and use violence against the public are often included, wealthy, and powerful.

Much of the recent past that we do know about, the two World Wars, the murders of the Kennedys and so many other fine people, the controlled demolition of 9/11, the recent murders of many young kids in Norway, in the light of this truth of a very ancient remote neuron reading and writing held tightly by a very wealthy minority, shows clearly that many of those who control remote neuron reading and writing, are not only terrible at stopping violence, but are apparently actively encouraging and funding violence against nonviolent people. For a people equipped with the ability to see inside houses and heads for centuries, it's obvious that they must be aware of thought-images of violent plans very far in advance and not only do they choose not to stop the violence, but many of those violent plans are created and/or remotely written by them.

Beyond this, that the public knows next to nothing about neuron reading and writing, or even the details of evolution, the history of science, and the future shows obviously that, these "neuron owners" are purposely making extreme efforts to keep the public extremely uneducated, and under-informed, by funding massive lies (like the wave theory for light, the theory of relativity, and the Big Bang expanding universe theory, etc.) and using violence, denial of D2B service, and other methods to coerce people who do care into silence.

"Isn't there anything we can do to stop people from remotely writing unwanted sounds, images, memories, muscle contractions, smells, and other sensations?"

Right now, in particular for D2B excluded, there is not much we can do, except to "fire back" in our mind. Because there are those who don't fire back that are easier targets, it may help those who do. But I think in the future, the majority of people who obviously must be opposed to remotely molesting people with unwanted neuron writing will be able to have much more control over the telecoms, the militaries of the various governments, and/or whoever controls the equipment of remote neuron writing. You can think of firing back as very quickly hanging up on thousands of unwanted phone calls. Certainly it may not be absolutely clear, in particular with sending images and sounds if the person receiving the data would find that image or sound annoying or helpful, but there are some obvious guidelines, for example, that nobody wants their muscles contracted- in particular if not to save them from some pain or imminent death, nobody wants to feel pain, most people do not want sounds or images when the sender(s) are not clearly and honestly identified, and so on. Neuron writing, and who gets to write to neurons, clearly is a much more important issue than neuron reading.

Even though I don't receive D2B, I'm all for it

Believe it or not, I actually am in favor of this technology, all the dust-sized cameras and microphones, and remote neuron reading and *consensual* remote neuron writing. In particular I would like to see D2B go public so we can all step up into this new age of communication with thought images and thought sounds. It's shocking and terrible that the wealthy keepers of remote neuron reading and writing have chosen to keep it a secret, providing D2B to only a minority of people, and then to strictly forbid them to talk about it. But then worst of all, to allow the abuse of those people they exclude from D2B service by letting terrible remote neuron writing be sent to their heads – without even telling them that remote neuron reading and writing is possible. Just an extremely unfair and unnecessary imbalance and segregation has happened.

In particular it would be very helpful and interesting to see the videos of all the famous people of the past and their thoughts. But also to see all the murderers – to know "who killed who?" would be important for many people. It saves a lot of time to not have to talk out loud using our muscles, but to just communicate using thought sounds and images. Initially it may feel like people are abusing your most private thoughts, but eventually you realize that all people have similar thoughts, that some are externally written, and that there is simply nothing beyond the "thought-screen" and "thought-audio" – that's all there is or will be for many centuries to come. Beyond that, it is better to be able to talk and think openly to each other about remote neuron reading and writing and the history of science.

Currently, we live in a dim age, where many people are being abused by remote neuron writing and secrecy is strictly enforced, but the picture of a future where remote neuron reading and writing is consensual and helpful to all people, and where everybody freely and openly talks about it seems inevitable to me.

Reading from and Writing to Neurons

By simply **reading** the values of neurons a person can: hear the sounds the ears hear, see the images the eyes see, hear the sounds the brain is thinking of (think of a song – to hear that song), and see the images the brain is thinking of (think of a food – to see that image). The signals of any sensation a brain receives- a touch, heat, pain, smells, tastes, etc. can be read and recorded. With neuron **writing**, you can actually write a sound to the ear neurons which is heard by the owner of the brain (as if it was an actual sound from outside their body), or write an image to the eye neurons of the brain which is seen by the owner of the brain (as if some object was actually in front of them), or write a sound to their thought audio neurons, which they hear in their thoughts (right where you hear any song you think of), or an image to their thought-screen neurons (exactly where you see images of food in your mind when deciding what you want to eat). In addition, since all muscles are connected to and controlled by nerves, writing to a neuron connected to a muscle can cause any muscle to contract. You probably have been a victim of remote neuron writing on many occasions and did not even realize it. Somebody you cared for may have actually been murdered ("galvanized") by remotely contracting the muscles that move their lungs, or their heart- all done remotely and invisibly with light particles and nanotechnology. It seems very likely that over the last seven hundred years a secret "remote particle murder" holocaust of massive scale has occurred.

Hundreds of years before now, it must have occurred to many smart wealthy people that if they can measure the light emitted from the electricity in neurons or even measure the electricity in neurons directly, then hearing and seeing thought might actually be possible. You can imagine that seeing and hearing thought has been a dream of many people for a long time, but the field of "telepathy" is usually, and no doubt purposely, associated with pseudoscience and fraud. People that do not get D2B need to look past the cover story and pseudoscience surrounding telepathy as some natural ability in a very few people, and to the likelihood that people long ago figured out how to see and hear thought using technology but kept it a secret for mostly greedy and to some extent for overly-paranoid reasons.

Imagine a person who owns a machine that only just hears thought, and then uses their advantage of hearing thought to maintain a monopoly on the technology of hearing thought. Wouldn't they be the most popular person on Earth? Wouldn't they instantly become millionaires? Wouldn't people flock to them and do all kinds of favors for them in order to hear the precious thoughts of their loved ones, and friends, their enemies and pets? Of course they would! Yes, without much doubt at all, a person that owns a machine that can hear thoughts would be very wealthy and popular.

There are some basic technology checkpoints that must have occurred in developing remote neuron reading and writing (electric storage, electric switches, etc.). **Direct** neuron reading and writing, is done by touching the neuron or body directly. **Remote** neuron reading and writing is done without having to touch the neuron directly. It may sound bizarre, but a human went public with direct neuron writing (also known as direct neuron activation) in 1678 (Swammerdam[1]) and with remote neuron writing in 1791 (Galvani[2]). In fact, in terms of public information, neuron writing came long before neuron reading. Perhaps the secret history is similar, with similar checkpoints, but with checkpoints occurring much earlier in the past. Direct neuron reading did not go public until 1875 (Caton[3]) and remote neuron reading did not go public until 2008 (Miyawaki, et al[4]) . Remotely reading and writing from and to neurons may be possible without any kind of other equipment, but it seems very likely that the creation of microscopic and nanoscopic devices brought the modern form of remote neuron reading and writing to perfection. Galvani put a scalpel on a frog leg nerve while an assistant cranked a distant spark generator. Light from the spark reached the scalpel and caused the frog muscle to twitch. Imagine now, that Galvani's scalpel is 1000x smaller and 1000x smarter but is still right there on the nerve. Like a modern RFID chip which may receive and send messages using light particles, such a device could

---

[1] John Joseph Fahie, "A History of Electric Telegraphy, to the Year 1837", E. & F. N. Spon, 1884.
http://books.google.com/books?id=0Mo3AAAAMAAJ
[2] Luigi Galvani, Elizabeth Licht, Robert Green, "Commentary on the Effect of Electricity on Muscular Motion", Waverly Press, 1953.
[3] Richard Caton, "The Electric Currents of the Brain", British Medical Journal, 1875, V2, p278.
http://www.bmj.com/content/2/765/257.full.pdf+html
[4] Miyawaki, Y., Uchida, H., Yamashita, O., Sato, M., Morito, Y., Tanabe, H. C., Sadato, N., Kamitani, Y. (2008). "Visual image reconstruction from human brain activity using a combination of multi-scale local image decoders. ", Neuron, 60, 5, 915-929.
http://www.cell.com/neuron/abstract/S0896-6273(08)00958-6

enter a body through the lung, then from there enter a blood vessel and go directly to many different cells. These devices can receive and send light particles with other similar devices floating and flying around (even through the skin)- sending the info read from and written to neuron cells all the way to and from some central data storage and command center.

It's hard to imagine that many millions of microscopic devices might be floating and flying around capturing images of us, remotely sending images to and from our thought-screens, and moving our muscles, but look under a ray of sun light or under a desk lamp– and you can see many tiny floating objects, like fibers. Imagine that each one of those might be some kind of communication, microphone, camera, laser, and neuron reading and writing device.

With remote neuron reading and writing, not only can pictures and sounds be sent to a person's eyes or ears, but any physical sensation can be remotely activated: seeing images, hearing sounds, smelling smells, feeling touches, itches, scratches, pains, pleasures, etc. can all be remotely sent. Remote neuron reading and writing brings to mind one of the great questions: How do we know that everything we experience is not remotely sent to our neurons and doesn't actually exist outside our body?

If you can send images remotely to the eyes and thought screen, and sounds remotely to the ears and thought-audio, then you can send a "Windows" interface, just like any computer, but no computer screen is needed – it is much more convenient – because the person can just see the windows in front of their eyes or in their mind, and hear the sounds without having to wear any "headphones". Think of the many windows open on your computer screen, but imagine them open just in front of your eyes. They don't necessarily block your view because they can be made semi-transparent – so you can still see what is going on around you behind the windows. The resolution of the human "eye screen" is very large, perhaps 100,000 x 100,000 dots (or neurons) and so windows could be very large, appearing far away in the distance. There must be "all immersive" windows, not semi-transparent, that allow a person to see and hear a movie as if they were actually in the movie since they only see and hear the movie, without seeing the external light entering the eye or hearing any external sounds. Movies we interact with while dreaming are examples of this phenomenon, but when we are sleeping we mostly don't realize that what we are seeing is just a movie.

Talking Openly about Neuron Reading and Writing

I may be inaccurate in some of these claims whether about direct-to-brain windows, or scientific theories. Because we excluded have to guess, probably some of our guesses will be wrong. It is always best not to take an absolutely certain view, for example running around yelling "people can hear our thoughts! Don't you get it?!", but instead to protect yourself from labels of crazy and to always remain calm, never express aggression and certainly not first strike violence. It's best to preface all explanations with "it may be", and "possibly", and to remember to periodically say "maybe I'm wrong – I don't know for sure". One way to explain is to say "isn't this idea of direct-to-brain windows interesting? Maybe in the future we will communicate through thought-audio, etc...", or "have you heard about remote neuron reading and writing? – it's really fascinating and many people don't even know that people have already figured out how to see eyes...", "wouldn't it be nice if we could get music beamed directly to our ear neurons? We wouldn't have to wear headphones", etc.

Chapter 3
Other Science Big Secrets, Lies, and Mistakes

There are some important science truths that I have learned that I think will help people tremendously in understanding direct-to-brain windows, remote neuron reading and writing, and the universe.

Because of the neuron lie and secrecy, many simple truths have not reached the public. In this book I am just going to summarize quickly these simple science truths. I am by no means an expert, and I do not claim to have all the answers and to be 100% accurate all the time, but many of these truths are just shockingly simple to understand and verify.

Light is made of material particles and is the basis of all matter

It seems very likely that light is made of material particles, and is not a transverse electromagnetic wave as is currently claimed. All galaxies, stars and planets are corpuscular, so it seems logical that light is made of corpuscular bodies too. In fact, all matter is probably made out of light particles, and the explanation is simple. When we light a candle, light particles from the candle and oxygen in the air are emitted into our eyes from the flame, and as time continues, the candle is made smaller in size. The only logical conclusion is that the light particles were in the candle (and oxygen) the entire time, and that the candle and all atoms are simply made of light particles – that the light particle (the "photon") is the basic atom of all matter. This is true even for the so-called "antimatter" because when a proton and antiproton collide, the result is not empty space, but, instead, of course, all matter is conserved and emitted as light particles and other larger fragmentary particles also made of light particles. Even "anti" particles are made of photons. All other particles besides photons (including atoms and molecules) are "composite particles". This includes all the sub-atomic particles (mesons, etc.) too- all made of light particles. The claim of relativity that light is "massless" is absurd, in particular when we realize that light is the basis of all matter, the most fundamental atom that all matter is made of. Perhaps even light particles are made of smaller particles.

It's useful to just quickly summarize a little history about the various explanations of light given to the public over the last few centuries. Long ago, around 56 BC, the Roman writer Lucretius described light as being made of atoms that move very fast[5]. The collapse of science and the rise of the Dark Ages delayed progress until around 1228 AD when the first president of the newly established Oxford University published a book theorizing that all matter is made of light[6]. So this idea of light being made of material particles, and being the basis of all matter, is not only not new, but is really pretty ancient. A more precise explanation of light did not reach the public until 1664 when René Descartes' book "Le Monde" was published, 18 years after his death. In "Le Monde" Descartes revives the theory that light is made of particles, comparing light to a ball, and is the first to describe the two major theories of light: the wave and corpuscular theory[7]. So Descartes is really a hero for comparing light to a ball. It's interesting to note that the earliest "wave" theories

[5] Titus Carus Lucretius, "T. Lucreti Cari De rerum natura libri sex, Volume 1", 1866, lines 176-229, p530
http://books.google.com/books?id=oiUTAAAAQAAJ
[6] Robert Grosseteste, tr: Clare C. Riedl, "On Light {De Luce}", 1942.
http://web.mit.edu/jwk/www/docs/Riedel%201942%20Grosseteste%20On%20Light.pdf
[7] Descartes, R. Le Monde ... Ou Le Traité De La Lumière Et Des Autres Objets Principaux Des Sens, Avec Un Discours De L'action Des Corps Et Un Autre Des Fièvres, Composez Selon Les Principes Du Même Auteur. Michel Bobin et Nic. le Gras, 1664, Chapters 13 and 14.
http://books.google.com/books?id=DHEPAAAAQAAJ
English translation:
Rene Descartes, Translated by Michael S. Mahoney, "The World or Treatise on Light", Chapters 13 and 14.
http://www.princeton.edu/~hos/mike/texts/descartes/world/worldfr.htm

are "corpuscular wave" theories; the wave is made of a material particle medium. The "wave" theory is probably more accurately called the "constant collision" theory, because in this theory light is transmitted by the constant collision of material particles that fill space, while the "corpuscular" theory is probably more accurately called the "rare collision" theory, because in this theory light is transmitted by particles that move through mostly empty space. It may be that inside dense objects like planets and stars light particles constantly collide and without moving much (like the wave theory describes), but when light particles reach the surface and empty space they rarely collide (like the corpuscular theory describes). In any case, a year later in 1665 Robert Hooke[8] and Francesco Grimaldi[9] firmly established the so-called wave theory for light, in which light is a *motion* through a medium of constantly colliding particles, and in 1672[10] Isaac Newton more clearly and firmly established the so-called "corpuscular" theory for light, in which light is made of material particles (corpuscles) that *move through* any medium. Newton compares the motion of light to a tennis ball, just as Descartes did. Many people recognize Isaac Newton as being the founder of the theory of gravitation, but few people know that Newton heroically established the corpuscular theory of light, which is perhaps of equal importance to the theory of gravitation, especially when we realize that material particles of light are probably

---

[8] Hooke, R. Micrographia: Or, Some Physiological Descriptions of Minute Bodies Made by Magnifying Glasses. With Observations and Inquiries Thereupon. printed for James Allestry, 1667, p56-57.
http://books.google.com/books?id=SgFMAAAAcAAJ

[9] P. Francesco Maria Grimaldo, "Physico-mathesis de lumine, coloribus, et iride", 1665.
http://books.google.com/books?id=sZE_AAAAcAAJ

[10] Isaac Newton, "A Letter of Mr. Isaac Newton ... containing his New Theory about Light and Colors", Feb 19, 1671/2, in English, c. 5,263 words, 13pp. Published in: Philosophical Transactions of the Royal Society, No. 80 (19 Feb. 1671/2), pp. 3075-3087.
http://www.newtonproject.sussex.ac.uk/view/texts/normalized/NATP00006

the basis of all matter. So initially everybody agreed that light was made of material particles, the disagreement was only how light is transmitted: by many collisions of particles as a wave like sound in air, or by particles moving mostly through empty space like a ball. The corpuscular (or mostly empty space, rare collision) theory of light, in my view the more logical and accurate theory, held popularity for about 100 years. This period was, in my opinion, a bright and progressive period for science. But all that changed, at least in terms of physics. The rebirth of logical and intuitive science of the 1700s collapsed in the early 1800s, and we still live in the Dim Era that resulted. In 1801, Thomas Young was the first person to publish the frequencies (and wavelengths or particle intervals) of various colors of light[11]. Which is a great achievement, but Young supported a wave theory for light. In the same paper, Young advanced the idea that rays of light "interfere" with each other, an effect for light similar to that for sound where two sounds will cancel to produce places of silence- the analogy being that two waves of light may cancel to produce places of darkness (which I examine in more detail later). Young (in 1817[12]) changed the traditional wave theory to claim that light is not a forward and backward ("longitudinal" or "point") wave as Descartes and then Hooke had described, but instead is a "transverse" wave; a wave with an amplitude. Young felt that a transverse wave could better explain the phenomenon of light interference. Augustin Fresnel (in 1821[13]) also sided

[11] Thomas Young, "The Bakerian Lecture: On the Theory of Light and Colours", Philosophical Transactions of the Royal Society of London (1776-1886),Volume 92, (1802), pp12-48.
http://books.google.com/books?id=-XAXAQAAMAAJ&pg=PA140
[12] "Letter from Dr. Young to M. Arago", Jan. 12, 1817, found in: Young, T., G. Peacock, and J. Leitch. Miscellaneous Works: Scientific Memoirs. Murray, 1855.
http://books.google.com/books?id=-XAXAQAAMAAJ&pg=PA380
[13] A. Fresnel, 'Considerations mecaniques sur la polarisation de la lumiere', Oeuvres, Vol. I, 629-49; Annales de chimie et de physique, Vol. XVII (cahier de juin 1821), 167 ff, p168.
http://books.google.com/books?id=lrc-AAAAcAAJ&pg=PA629

with this transverse wave theory. Sadly, this (in my opinion) unlikely wave or "undulatory" theory replaced the more logical corpuscular theory in popularity. As an aside, many people probably don't know that Young is credited with being the first, in 1807[14], to formally apply the word "energy" (the "currency" of scientific theories from the 1800s on) to Leibniz's "vis-viva" ("living force") quantity mv^2 of 1695[15]. Moving on with the story, James Clerk Maxwell, in 1864 (and 1873), accepted and adjusted Young's transverse wave theory by supposing that light is made of not one transverse sine wave in an aether, but instead, is made of two waves: one an electric wave, and the other a magnetic wave, both positioned at 90 degrees to each other.[16,17] And this (in my humble view) very unlikely "electromagnetic" interpretation has, shockingly, held to this sad day. For example, the spectrum of light is still called the "electromagnetic spectrum". But the corpuscular side fought back and in 1881 Albert Michelson, a "Master" in the US Navy, reported an experiment that showed that the speed of light is unchanged relative to the motion of the Earth around the Sun through the supposed aether medium. If there is a medium for light, and it is stationary relative to the motion of the rest of all the matter in the universe, then this motion should make the light waves take less time in the direction of the Earth's motion around the Sun, but this wasn't found. Michelson

[14] Young, T. A Course of Lectures on Natural Philosophy and the Mechanical Arts. Johnson, 1807. A Course of Lectures on Natural Philosophy and the Mechanical Arts., p78.
http://books.google.com/books?id=YPRZAAAAYAAJ&pg=PA78
[15] Gottfried Leibniz, "Specimen Dynamicum" (1695).
http://books.google.com/books?id=0je_DN18UkoC&pg=PA315
English translation: L. E. Loemker, "Philosophical Papers and Letters", (1976), pp.435-452.
[16] James Clerk Maxwell, "A Dynamical Theory of the Electromagnetic Field", Royal Society Transactions, Vol. 155, 1865, p. 459-512.
http://books.google.com/books?id=xVNFAAAAcAAJ&pg=PA459
[17] James Clerk Maxwell, "A treatise on electricity and magnetism.", 2 vol., 1st ed, Oxford, 1881, p383-398.
http://books.google.com/books?id=gmQSAAAAIAAJ&pg=PA383

was even so bold as to write that "The result of the hypothesis of a stationary ether is thus shown to be incorrect, and the necessary conclusion follows that the hypothesis is erroneous.".[18] And then giving a final stamp of approval from none other than the D2BW specialist himself Alexander Graham Bell writing: "In conclusion, I take this opportunity to thank Mr. A. Graham Bell, who has provided the means for carrying out this work". Without any aether medium, it is hard to imagine light as a wave. This made the corpuscular theory for light appear to be the more accurate interpretation. But like the Empire, the wave camp struck back. To explain this experiment of 1881 (and the mysteriously more popularized experiment of 1887[19]) and to try to save the theory of light as a wave that moves through an aether medium that fills the universe, George FitzGerald, in 1889[20] developed a theory that matter contracts just enough to account for the absence of any apparent difference in the speed of light that would be due to the movement of the Earth through a stationary aether. In 1892 (and 1899) Hendrik Lorentz threw his support behind this theory and expanded it by saying that not only does matter contract in the direction of motion just enough to offset it's motion through a stationary aether, but that even time itself can be contracted or dilated depending on the motion of matter[21,22]. Lorentz

---

[18] Albert A. Michelson, "The relative motion of the Earth and the Luminiferous ether", The American Journal of Science, Volume 122, 1881, p120.
http://books.google.com/books?id=S_kQAAAAIAAJ&pg=PA120
[19] Albert A. Michelson and Edward W. Morley, "On the Relative Motion of the Earth and the Luminiferous Ether", American Journal of Science, s3, v34, Num 203, 11/1887, p333.
http://books.google.com/books?id=0_kQAAAAIAAJ&pg=PA333
[20] George FitzGerald, "The Ether and the Earth's Atmosphere.", Science, Vol 13, Num 328, 1889, p390.
http://books.google.com/books?id=8IQCAAAAYAAJ&pg=PA378
[21] H. A. Lorentz, "The Relative Motion of the earth and the Ether", Konink. Akademie van Wetenschappen te Amsterdam, Verslagen van der gewone Vergaderingen der Wis- en Natuurkundige Afdeeling, 1892, 1:74 ff; also in H. A. Lorentz, Collected Papers (The Hague: Martinus

theorized that, instead of there being one time for all of the universe, each piece of matter has it's own time. Most of us can probably recognize these theories as being doubtful. Michelson rejected these unlikely explanations even as late as 1927 writing: "Lorentz and Fitzgerald have proposed a possible solution of the null effect of the Michelson-Morley experiment by assuming a contraction in the material of the support for the interferometer just sufficient to compensate for the theoretical difference in path. Such a hypothesis seems rather artificial, and it of course implies that such contractions are independent of the elastic properties of the material."[23]. The next big development happened in 1900 when Max Planck partially revived the particle theory for light by theorizing that objects emit light in individual units he called "energieelements" but which were later called "quanta". Planck created a simple equation $E=h\nu$ where E is an Energieelement, h is a constant and ν is the frequency of emitted light[24]. In my view, Planck's "quantum" theory will be remembered

Nijhoff, 1937), vol 4., pp219-223.
http://books.google.com/books?id=8Q9WAAAAMAAJ
[22] H. A. Lorentz, "Théorie simplifiée des phenomènes electriques et optiques dans des corps en mouvement.", Traduit de Versl. K. Akad. Wetensch. Amsterdam 7, 507, 1899.
"Simplified Theory of Electrical and Optical Phenomena in Moving Systems", Proceedings of the Royal Netherlands Academy of Arts and Sciences, 1899 1: 427-442.
http://en.wikisource.org/wiki/Simplified_Theory_of_Electrical_and_Opt ical_Phenomena_in_Moving_Systems
[23] Albert Michelson, "Studies in Optics", Chicago University Press, 1927, p156.
[24] M. Planck, "Zur Theorie des Gesetzes der Energieverteilung in Normalspektrum," Verhandlungen der Deutsches Physikalisches Gesellschaft 2 (1900), pp. 237-245.
http://books.google.com/books?id=zYYMAAAAYAAJ&pg=PA237
and "Uber das Gesetz der Energieverteilung im Normalspectrum", Annalen Der Physik, 4 (1901), p553-563.
http://books.google.com/books?id=j6AqAAAAYAAJ&pg=PA553
English translation:
Max Planck, Alexander Ogg, "On the Law of Distribution of Energy in the Normal Spectrum", 1903.
http://theochem.kuchem.kyoto-u.ac.jp/Ando/planck1901.pdf

mostly for partially reviving a particle theory for light, because most of the "energy" work of the 19th and 20th century, like the theories of relativity, are abstract and overvalued, especially when looking at the secret and more practical development of robots, rockets, transmutation, RNRAW, and nanoscale particle devices. Albert Einstein was an early supporter of Planck's quantum theory, and applied it to describe the photoelectric effect famously in March of 1905[25]. So Einstein initially appeared to be a supporter of a corpuscular theory for light. I had said that it is hard to imagine light being a wave without an aether medium, but Albert Einstein could and did. In June of that same year[26] Einstein published his famous "Special Theory of Relativity" which adopted the far-fetched aether-saving theories of FitzGerald and Lorentz that matter and time contract depending on the speed of a material object relative to a stationary aether, but in an ironic twist, and perhaps as a compromise with the (by that time) nearly extinct corpuscular school, Einstein rejected an aether medium for light as being "superfluous", choosing instead to base motion relative to an observer. In this view the mass of an object increases the faster it moves (gaining mass from some unknown source), and time, as experienced by the object, slows down the faster it moves relative to (presumably) the rest of the matter in the universe. In September of 1905 Einstein publishes

---

[25] A. Einstein, "Über einen die Erzeugung und Verwandlung des Lichtes betreffenden heuristischen Gesichtspunkt", Annalen der Physik (ser. 4), 17, 132-148.
http://www.physik.uni-augsburg.de/annalen/history/einstein-papers/1905_17_132-148.pdf
English translation: "On a Heuristic Point of View Concerning the Production and Transformation of Light"
http://users.physik.fu-berlin.de/~kleinert/files/eins_lq.pdf
[26] A. Einstein, "Elektrodynamik bewegter Körper", Annalen der Physik (ser. 4), 17, 1905, 891-921.
http://www.physik.uni-augsburg.de/annalen/history/einstein-papers/1905_17_891-921.pdf
"On the Electrodynamics of Moving Bodies"
http://users.physik.fu-berlin.de/~kleinert/files/eins_specrel.pdf

his famous $E=mc^2$ equation (originally $m=L/c^2$)[27], which, to Einstein's credit at least suggests the theory that all matter might be made of light, but, to me at least, implies that matter and motion can be converted into each other, and that mass has some inherent motion, both of which I doubt. Then 10 years later in 1913[28], as if matter and time contraction and dilation depending on motion is not unlikely enough, Einstein (with help from Marcel Grossman) expanded this theory and adopted the radical and unlikely surface topology (so called "non-Euclidean") geometry (which I describe below), to try to explain the movement of matter in the universe in his famous "General Theory of Relativity". It is, in my humble opinion, and with all due respect, very unlikely that a restricted surface geometry applies to the movements of matter through time in the universe. But as radical, extreme, and unlikely as applying a surface geometry to the universe may seem to me, this explanation is currently the most popular. In 1922, Louis De Broglie joined Planck's $E=hv$ with Einstein's $E=mc^2$ to solve for the mass of what he called "atomes de lumière" (atoms of light). The mass of the light quantum is now described as it's "rest mass" because of the doubtful (and no doubt deliberately dishonest) claim that motion changes a body's mass. De Broglie didn't give an

[27] A. Einstein, "Ist die Trägheit eines Körpers von *seinem Energieinhalt* abhängig?", Annalen der Physik (ser. 4), 18, 639–641.
http://www.physik.uni-augsburg.de/annalen/history/einstein-papers/1905_18_639-641.pdf
English transation: "Does the Inertia of a Body Depend upon its Energy Content?"
http://users.physik.fu-berlin.de/~kleinert/files/e_mc2.pdf
[28] A. Einstein, M. Grossmann, "Entwurf einer verallgemeinerten Relativitätstheorie und eine Theorie der Gravitation. I. Physikalischer Teil von A. Einstein II. Mathematischer Teil von M. Grossmann", Zeitschrift für Mathematik und Physik, 62, 1913, 225–244, 245–261.
English translation: "Outline of a Generalized Theory of Relativity and of a Theory of Gravitation. I. Physical Part by A. Einstein II. Mathematical Part by M. Grossmann". In A. Einstein, Edited: M. Klein, et al, "The Collected Papers of Albert Einstein: Vol 4, The Swiss years: writings, 1912-1914", 1995.

exact number but simply supposed that light atoms have a "masse très faible" (very low mass). This raises the startling truth, although I haven't scoured the archives, that no actual mass of a light particle in terms of grams has ever been made public. In 1909, Jean Perrin described the mass of the electron, taken to be 1000 times less than Avogadro's number, as 0.805×10^{-27} grams, so perhaps the actual mass of light particles in terms of grams is 1000 times smaller around 1×10^{-30} grams. It also raises the point that, really, the light particle is perhaps a better "base" unit of mass than an atom is. Perhaps we should describe small masses in terms of "number of light particles". Finally, around 1927 Arthur Compton coined the name "photon" for the "quantum of light"[29] (note that the term "photon" is currently not defined as a material particle, as I am saying it should be, so perhaps it should be officially redefined, or a new word should be created like "luxon" or "lighton" for the ancient interpretation of light as made of material particles). So that brings us to the present time where chaos still reigns and truth is still viewed as disease.

The Universe is infinite in size and age, the Big Bang and Expanding Universe Theories are probably false

It seems very likely that the universe is not expanding, and that there was no "Big Bang", as hard as that is to accept or understand for many people. The best reason for this that I can give is simply that there must be many galaxies so far away that not one particle of light emitted from them can reach us in the tiny part of the universe that we occupy. Any so-called background "radiation" (why do they never say background "light particles"?) can only be light particles from light sources that are near enough to reach us. When we build a bigger

---

[29] A. Compton, "X-rays as a branch of optics", 12/12/1927.
http://nobelprize.org/nobel_prizes/physics/laureates/1927/compton-lecture.pdf

telescope and then receive light from more distance sources, will we then say that the universe just got bigger and older? Given this simple truth, it seems likely to me that the universe must be infinite in size, age and scale. Either theory, "Big Bang" or "Infinite Universe" are difficult to comprehend.

I find the "Infinite Universe" view, far more interesting and likely. In this view the universe is infinite is size and age, with no beginning or end in space or time. The universe is probably infinite in scale too. Stars, galaxies or even galactic clusters might be only light particles at some larger scale, and likewise, light particles at our scale, might be stars, galaxies or galactic clusters on some smaller scale. At any scale, light particles just bounce around forming different structures. In this sense, nothing really changes when a body is born, lives or dies- the same exact light particles just continue on as usual in their unknown course.

Beyond this idea that there must be matter so far away that not one particle from it can reach us, is the historical data involved with the expanding universe Big Bang Theory – which many people have never been shown (for example, how many of you have ever seen figure 3.1 before?). Of course, because of the big neuron lie and many decades of silence about remote neuron reading and writing, you can be sure that corruption has played a major role in what "official" science theories are told to the public. The nicest thing that can be said is that this claim of a "red shift" is simply a mistake, but the evidence for D2B and the obvious and simple nature of the Bragg equation for gratings (which I will explain later), imply that this claim of Doppler shift was a very corrupt and intentionally dishonest claim that the authors knew was false, and that is mass marketed to the public like so many other famous claims (Oswald killed JFK, Sirhan killed RFK, 9/11 was 19 hijackers, etc.) while any other theory, in particular, the truth, is mostly banned from the market.

In the 1800s it was thought that the changing position of spectral lines could be used to measure the Doppler shift of celestial objects - how the frequency of light from a light source changes because of the relative motion of the light source toward or away from the viewer. Slipher was the first to measure the supposed Doppler shift of other galaxies by comparing the positions of two common absorption lines attributed to calcium, publishing it (without any photos) in 1913[30]. To my knowledge, the first actual image of these shifted calcium absorption lines was not made public until Milton Humason published some in a famous photo in 1936[31] (see figure 3.1). Looking at this photo more closely we see the visible spectrum of five different galaxies (although only in black and white). The two calcium absorption lines are apparent on the left (blue) end of the spectrum. Now notice how the overall width of the spectrum becomes smaller for each more distant light source. A result of this reduction in spectrum size is that the calcium lines appear to be closer to the center of the spectrum- just as part of the red end of the spectrum appears to shift in the blue direction (to the left). Are we to believe that part of this galaxy (the blue end) is moving rapidly away from us, but another part (the red end) is moving rapidly toward us?!

[30] Slipher, V. M., "The radial velocity of the Andromeda Nebula", Lowell Observatory Bulletin, vol. 1, pp.56-57.
http://adsabs.harvard.edu/full/1913LowOB...2...56S.
[31] Humason, M. L., "The Apparent Radial Velocities of 100 Extra-Galactic Nebulae", Astrophysical Journal, vol. 83, p.10, Jan 1936.
http://articles.adsabs.harvard.edu//full/1936ApJ....83...10H/0000011.000.html

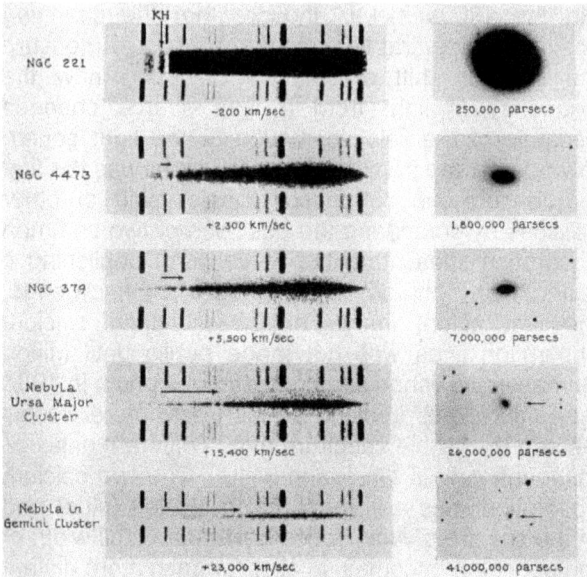

Figure 3.1. Image from a 1936 Milton Humason paper. Notice how the spectrum size gets smaller for each smaller light source, which in turn moves the two absorption lines closer to the center of the spectrum. In the bottom spectrum, is part of the galaxy coming rapidly at us because the right (red) end is blue-shifted?

As obvious as this observation is, no major astronomers have publicly stated "the claimed red-shift of these 1936 spectra is not accurate because the spectra have different widths". It seems very likely to me that most of this shifting is due to the different sizes of the spectra. But in addition, some of this shifting may be the result of the simple Bragg equation (as I will explain later), because for each particular frequency of light reflected from a grating, there is a specific angle of incidence required. As a result, from simple trigonometry, that angle must occur at different positions along the grating for two light sources of different distance (see figure 3.7). So this shifting of the calcium lines and the spectrum in general is not a shifting of *frequency* but most likely a shifting of spectral line *position* only because the light sources are different in size and distance. The 1936 spectra have an obvious problem with *scale*- the size of each spectrum is not equal. When we stretch out the spectra to make them all equal in width, most of the apparent shift disappears.

Then twenty years later (after World War 2, in 1956), Humason, this time with astronomers Mayall and Sandage, mysteriously revisit the "red-shift" claim after 20 years of silence, and publish a second infamous photo (see figure 3.2). Unlike the first 1936 photo, not only are the calcium absorption lines in this photo much more difficult to see, but their supposed position relative to the rest of the spectrum is far from their appearance for the galaxies in the 1936 photo. The only other photos of the full visible spectrum with the calcium absorption lines that I have ever seen are derived (apparently) from these 1956 spectrographs. These are the famous 5 spectra that are printed in some (even modern) Astronomy textbooks and most likely the spectra shown in Carl Sagan's 1984 PBS television series "Cosmos" (see my paper on the web to see that image[32]). The "Cosmos" photo has apparently been "colorized" because there is no way that the full spectrum of a galaxy can have no color "red" in it. But if we then "re-colorize" it, making the right-most part red, then it is apparent that the frequencies of this spectrum do not match the frequencies of the reference spectrum (Hydrogen), for which red is farther away to the right. So here again, we are confronted with the truth that spectral lines change position depending on the size and distance of the light source, just as the Bragg equation requires.

---

[32] Ted Huntington, "Spectral line position depends on distance of light source - Bragg Equation Effect", 05/10/2011.
http://tedhuntington.com/paper_Bragg.htm

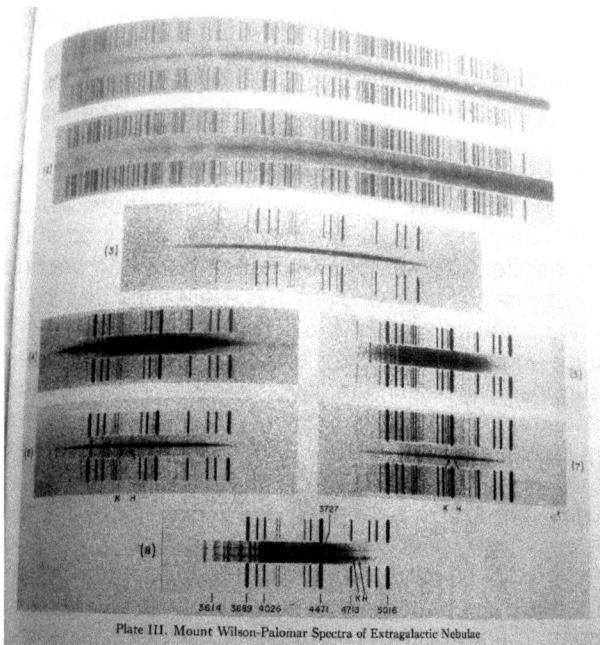

Plate III. Mount Wilson-Palomar Spectra of Extragalactic Nebulae

Figure 3.2. Image from Humason, Mayall and Sandage's 1956 paper, where a much bigger shift is claimed. I see the two absorption lines in the top two spectra in their usual position, but in the bottom three, a pair of similarly clear absorption lines are not apparent. If this was a pregnancy test, I would want another.

If I am wrong: where are all the color photos like Humason's black and white 1936 (figure 3.1) and 1956 (figure 3.2) images showing the public the shifted calcium absorption lines? I have yet to see any other images of the supposed shifted calcium absorption lines other than those two from Humason in 1936 and 1956. Even a modern astronomy book like "Astronomy: a beginner's guide to the universe"[33] (2004) includes the same exact above spectra images published back in 1956- is there no more modern or even *any* other image of the shifted

---

[33] Chaisseon, McMillan, "Astronomy, A Beginner's Guide to the Universe", 2004, p422.

calcium absorption lines? This suspicious photo is becoming iconic and representative of fascism (facilitated by D2BW) in the United States, like the altered "Life" magazine cover photo of Lee Harvey Oswald[34], the "fall-guy" for the actual JFK murderer. Perhaps some amateur or even professional astronomers will be able to provide more spectra of those same galaxies Humason published either showing or not showing calcium absorption line shift.

Note that whenever anybody doubts the expanding universe, insiders instantly present the only other alternative as being the "steady-state" universe theory, an unlikely theory (with all due respect) created by the astronomer Fred Hoyle in 1948[35], in which matter is constantly being created (from empty space) and destroyed in the universe. The simple answer to a "continuous creation" or "steady-state" theory is that there is no need to create or destroy matter when all matter is made of light particles that are never created or destroyed but simply move around in the space of the universe. Again, D2BW consumers all know this, but are paid and/or coerced by their D2BW dealer to "play dumb" and pretend to be completely unaware of a "matter was always here and is never created or destroyed" universe theory.

Globular clusters are made by advanced living objects and are the inevitable result of natural selection of the best adapted species- our future is to try to build a globular cluster

Like the simple truth about light being a material particle, and the immense value of remote neuron reading and writing, and bipedal robots (which I will talk about soon), I have to scratch my head and wonder why the popular people in science and education have remained silent for centuries, not

---

[34] "Life" (Feb. 21, 1964).
[35] Hoyle, F., "A New Model for the Expanding Universe", Monthly Notices of the Royal Astronomical Society, Vol. 108, p.372.
http://adsabs.harvard.edu/full/1948MNRAS.108..372H

even entertaining the public with the idea that globular clusters are made by living objects. Globular clusters (see figure 3.3) are groups of stars found around all spiral galaxies. Clearly, we are going to build cities on the moon of Earth, on Mars and the other moons and planets. Our population is going to continue to grow and grow in number if we are successful. And of course, we are going to send ships to and build cities around the other stars. Gravitation provides a very simple way to pull stars closer together- by simply using another mass to pull them along in any direction wanted (although the pulling object would have to be very massive, the better method is probably using many coordinated smaller masses). Given this truth, when we go to those other stars and pull them closer, to save fuel and time in moving back and forth between them, we would look a lot like a globular cluster. Just like bacteria and humans, probably living objects around globular clusters do exactly the same thing, convert matter into more of them. Like us, living objects around the stars in globular clusters probably have a strong desire to explore the universe.

Figure 3.3. Globular cluster M15 in the Milky Way Galaxy: one end product in the survival of the fittest. Our future is to build a globular cluster.

The big picture: Nebula to Spiral to Globular

An interesting related note to this truth about globular clusters is that there is a simple pattern in the universe (see figure 3.4): light particles move to empty spaces, become trapped with each other, growing to form atoms. Eventually as more light particles become trapped, a nebula is formed, then a spiral galaxy. Living objects evolve around stars in the spiral galaxy and start to convert matter into more of them, pulling stars close together as they colonize them. Eventually the galaxy is all globular and is what is being called an "elliptical galaxy" (although I think "globular galaxy" is probably a better term). It may be that a globular galaxy can exist for a long time by consuming other galaxies or existing on incoming light particles. Natural selection operates everywhere in the universe, and galaxies and globular clusters may be viewed like fish in an ocean of space- the most adapted survive, the less adapted disintegrate. If most globular galaxies emit more light than they take in, then they would ultimately emit the light particles stored in stars and be constantly reduced in size until they were almost empty space again.

This basic pattern seems logical to me, but no major astronomers or other scientists are saying this. They tell the public that globular clusters are made of "generation two" stars. They say nothing about a globular cluster being a massive matter conversion center made by very adapted living objects where any and all available matter is pulled in, consumed, and changed into more of them. But if they are D2B consumers, how could they not know such a simple truth?

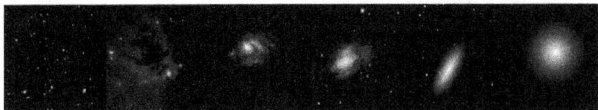

Figure 3.4. Stages in the evolution of galaxies (from left to right): Empty space, LDN1622, M8, NGC3521, NGC3115, M87.

Where are the bipedal (two-leg walking) artificial muscle robots that can shop, cook, and clean?

First the bipedal robots. Are we to believe that people who mass produce personal computers could not build walking robots by now? Come now. Knowing that many people probably figured out how to see, hear and send images and sounds to and from brains long ago, adds to the doubt. And beyond that, that the simple contraction of muscle, which the majority of animals do all the time, has not been achieved in an "artificial muscle" by the many thousands of chemists in multibillion dollar governments, universities and companies just like synthetic rubber and fabric was? The much more likely truth is that these inventions (very smart and fast moving bipedal robots and even light-weight artificial muscle robots) were probably developed many years ago, but those who control technology on Earth have somehow, out of unjustified fear, and perhaps greed, chosen to not allow these devices to enter the public. Think of how many lives could be saved by robot vehicle drivers with millisecond reaction times as opposed to the many mistake making humans that drive. Think of how much easier life would be for the average middle income person to have a robot to cook and clean for them? Perhaps part of the problem is the fear many old-world people have that robots will somehow kill off humans. This classic scenario is impossible in my mind, because, like particle devices, there is always a kind of stale-mate between two opposing sides. It seems very likely to me that for a long time into the future robots will always be working for humans – cooking, cleaning, etc. and will be the first to reach the other stars from Earth.

Any atoms can be fissioned and fusioned

Any atoms can be split (fissioned) and fused together (fusioned). The public isn't being told this, and so most people don't know this. I didn't know this until I found two papers from 1950.[36,37] This process of changing one atom into another is traditionally called "transmutation". Apparently atoms can be broken down and built up using particle accelerators. In the first paper the authors explain that even atoms in the middle of the Periodic Table can be split apart. In the second paper, the authors explain how they build up larger mass atoms by colliding material with carbon ions. The traditional public story is that the quantity of atoms changed is so small for the quantity of electricity used, that the process is mostly useless. But it is obvious to me that transmutation is extremely important. Most moons and planets in this star system have very little free oxygen and hydrogen. The idea of transporting oxygen and water from Earth to other planets is very unlikely. Since all atoms are made of light particles, the goal is to make use of the tons of atoms right there on those distant moons and planets by converting the common atoms (like Silicon and Iron) into more useful atoms we need like Hydrogen and Oxygen. So just like the simple idea of all matter being made of light, of the value of remote neuron writing and artificial muscles, so it is with transmutation – just nearly absolute silence. The silence is one of the biggest giveaways. Such ripe and valuable veins of scientific research would normally be massively explored and openly discussed as an important future goal. And most likely they have, but with a strict and most likely very harshly punished code of silence. Probably by now, just like remote neuron reading and writing nanotechnology, very efficient conversion of iron and other common atoms into a wide variety of more useful atoms has probably been achieved – but only

[36] Roger E. Batzel and Glenn T. Seaborg, "Fission of Medium Weight Elements", Phys. Rev. 79, 528 (1950).
http://prola.aps.org/abstract/PR/v79/i3/p528_1

[37] J. F. Miller, J. G. Hamilton, T. M. Putnam, H. R. Haymond, and G. B. Rossi, "Acceleration of Stripped C12 and C13 Nuclei in the Cyclotron", Phys. Rev. 80, 486 (1950).
http://prola.aps.org/abstract/PR/v80/i3/p486_1

for a group of perhaps a few million people that are in on the secrets.

Matter and motion are apparently completely separate and cannot be exchanged

The traditional view is that "energy" is the base form of all matter, and both matter and motion can be converted into each other. This is implied by any equation of energy ($E=1/2mv^2$, or $E=mc^2$, etc.). But simple logic shows that matter is clearly conserved, and that motion is clearly conserved, but matter cannot be converted into motion, and motion cannot be converted into matter. Conservation of matter is widely accepted as true by people, but not conservation of motion for some reason. The same is true for momentum ($p=mv$). Motion can be transferred from one piece of matter to another, but it seems unlikely to me that new matter or motion can ever be created or destroyed.

There is a comforting view that this possible truth provides: when a person you love dies, they are always still here, no part of them has disappeared from the universe; the light particles that they were made of simply move on to continue their mysterious and unknowable journey in the universe.

Diffraction, interference, polarization, refraction, and double refraction are all actually particle reflection.

Diffraction

In 1665[38] Francesco Grimaldi invented the term "diffraction" to describe the way that when light is passed through two holes, some light appears to "bend" and be seen outside the cone of light (see figure 3.5). The problem with this explanation is that Grimaldi never takes into account reflection of light

---

[38] P. Francesco Maria Grimaldo, "Physico-mathesis de lumine, coloribus, et iride", 1665.
http://books.google.com/books?id=sZE_AAAAcAAJ

from the side of the hole which definitely explains how light can appear outside of the cone of light.

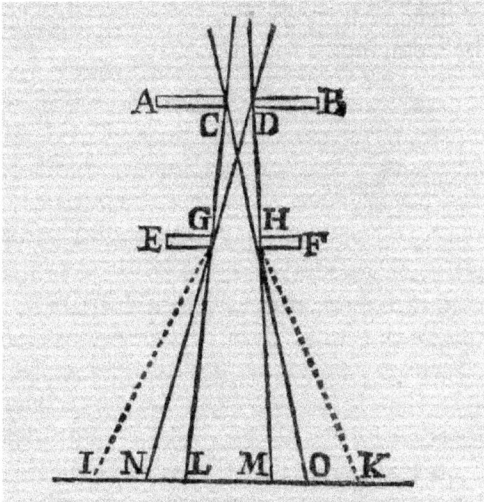

Figure 3.5. Image from Francesco Grimaldi's book "De Iride" showing what Grimaldi claims is a new property of light he calls "diffraction" But IN and OK are probably the result of particle reflection off the inside surface- think of a ray coming from above, reflecting at H and landing at I- lines of reflection are not drawn but must exist.

On 11/11/1912 (the date famous papers are published may sometimes have secret significance) William Lawrence Bragg showed that so-called "diffraction" patterns, the way the different frequencies of light spread out into the familiar color spectrum when light collides with a grating (a plate of glass with many equally spaced carved lines), can be explained as a particle reflection phenomenon (see figure 3.6)[39]. However, apparently many people have (no doubt purposely) failed to show that the Bragg equation can also explain any so-called

[39] Bragg, W.L. "The Diffraction of Short Electromagnetic Waves by a Crystal.", Proceedings of the Cambridge Philosophical Society, 1913: 17, pp. 43-57.
http://ulsfmovie.org/docs_pd/Bragg_William_Lawrence_19121111.pdf

diffraction phenomenon. In Bragg's interpretation, particles reflect off the regular rows of atoms, just like the regular grooves of a grating, at specific frequencies defined by the angle of incidence and the space between atoms (or grooves).

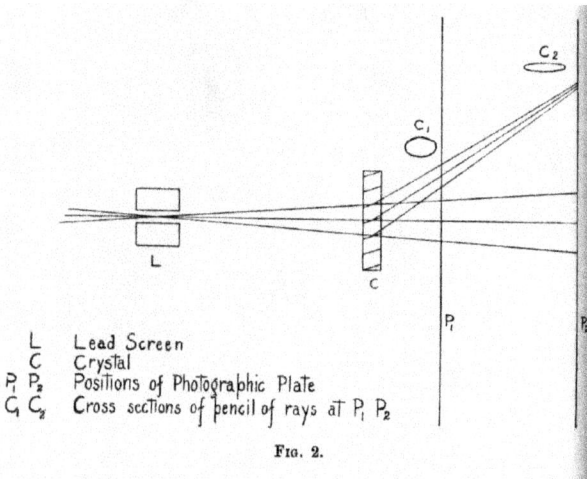

L   Lead Screen
C   Crystal
P_1 P_2 Positions of Photographic Plate
C_1 C_2 Cross sections of pencil of rays at P_1 P_2

Fig. 2.

Figure 3.6. Image from William Lawrence Bragg's 1912 paper "The Diffraction of Short Electromagnetic Waves by a Crystal" showing how diffraction can be interpreted as a particle reflection.

Bragg made use of a simple equation $n\lambda=2D\sin\theta$ that he attributes to Schuster, but which is now called the "Bragg equation", to describe this property of gratings. In this equation n is the order of the resulting spectrum, λ the resulting reflected "wavelength" (or equivalently "particle spacial interval" or "distance between particles"), D the space between atoms (or grating grooves), and θ the angle the light makes with the atomic plane (or grating groove). So to create a particular frequency of light particles in the reflected spectrum, a precise angle and grating groove spacing is required. For example, green light (570 nm) on a grating with 1um between grooves requires an angle of incidence (with the groove) of 16°, but to reflect blue light (380 nm), a smaller angle of incidence, 11°, is required. So you can see that each different frequency of light in the colorful spectrum from a single light source comes from a different part of the grating.

Because of this Bragg equation, simple trigonometry shows (see figure 3.7) that if two light sources are at different distances, the position on the grating of any particular frequency of light must be located at different places relative to the center.

A simple demonstration of this effect is to look through a $10 sheet of plastic grating you can buy on the Internet while walking toward a light. You can see that relative to the center, as you get closer to the light, the spectral lines move closer to the center (are "blue-shifted"), as you move farther away from the light the spectral lines move farther from the center (are "red shifted"). Another example is looking through one of those low-cost plastic "black tube" diffraction gratings at the Sun (only for a second so you don't hurt your eye). In order to see the colorful image of the Sun (not just the spectrum from ambient light- but the actual circular image of the bright Sun) in the black tube you need to look "next to the Sun" because when looking through the tube directly at the Sun, none of the rays of sunlight directly from the Sun can "make the angle" necessary (with the plastic grating near the eye) to reflect the frequencies in the visible spectrum.

Bragg equation: $n\lambda = 2D\sin(\theta)$
θ is the angle of incidence required for any particular color

For a grating with D=1000 lines/mm:
$\theta_{red\ (760\ nm)} = 22°$ but $\theta_{blue(360\ nm)} = 11°$

γ = Groove angle to perpendicular
$\alpha = 90 - (\theta + \gamma)$

$X_{s1} = Y_{s2}/\tan(\alpha)$
$X_{s2} = Y_{s1}/\tan(\alpha)$
$X_{s1}Y_{s2} = X_{s2}Y_{s1}$

The location of the spectral line
changes depending on the
distance to the light source.

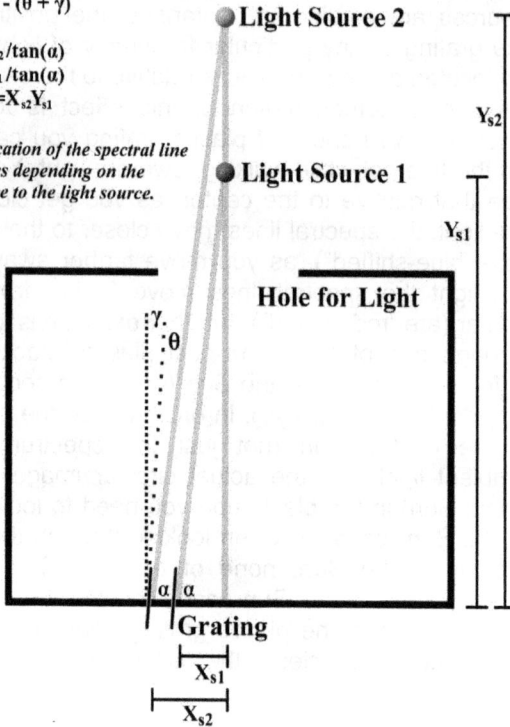

Figure 3.7. Simple trigonometry shows that two light sources at different distances cannot achieve the same angle with the grating grooves at the same location on a diffraction grating. This may explain some of the changing position of spectral lines from light sources of different distance. This effect is shown for two identical lamps at different distances in an image on the back cover of this book.

The details of the particle reflection explanation for "diffraction" that are described by the Bragg equation are more complicated than you might expect for so simple an equation. Many different frequencies of light reflect off each grating groove surface, in very small quantity for each groove. But for many angles of incidence, there is a resonance between multiple beams over a group of grooves which reflect specific frequencies of light particles in a single specific direction. Many light sources, like stars, emit light particles in many directions from a central sphere, and this creates a progressively increasing angle of incidence from the center of the grating to the outer edge. All the frequencies of light that are reflected off a grating groove wall, for some specific angle of incidence, reflect in the same direction, but there is a stronger intensity (more light particles) over a few successive grooves, for one particular interval (wavelength) of light in one particular direction which is related to the angle of incidence and space between grooves. Other rays of light in between these resonant rays pass through without reflecting in the space between each groove. You can imagine the horizontal space between gratings as being rotated 90 degrees to become an identical vertical interval between light particles. Particles with a space between them that is related to the space between grating grooves, reflect off of a small group of successive grooves at the same time, in the same direction, to produce a signal strong enough to see. Not many light rays reflect this way, because the space where the angle of incidence and groove spacing align for any frequency is very small, so the spectrum of light is produced by a very small fraction of rays from the source light.

In a prism the color spectrum may also be the result of the angle "refracted" light makes with multiples of the regular spacing between silicon atoms in glass.

Bragg showed this equation to be true for x-rays, but it is obvious that this equation applies to all particles of matter (protons, atoms, molecules, etc.), and then not only just for x-ray frequencies, but for all frequencies of particles even radio-spaced gratings, and with any size particles- even sand grains and larger objects must produce the same phenomenon.

As an interesting footnote, I recently found that the University of London professor Herbert Dingle had recognized this change in spectral line position with light source distance based on the simple Bragg or "grating" equation over 50 years ago in a 1960 work.[40]

Interference

In his famous 1801 paper[41], Thomas Young formally introduced the claim that light waves interfere in a way similar to sound waves; where the waves can add or subtract from each other. The most traditional experiment used to demonstrate light interference is the "double-slit" experiment: a beam of light is sent through a double-slit which, when done correctly, will produce numerous lines, the dark areas being where light waves are supposed to "cancel out". But similar to Grimaldi's diffraction claim, reflection of particles of light off the inner sides of the slit (and off each other) are never accounted for. What many people aren't shown is that, a single-slit can produce similar lines, which implies that this distribution of light is a reflection phenomenon. In addition, we don't see light interference from two identical visible light sources (like two lamps in a room) without any reflecting slit apparatus. A simple experiment may show that light interference of two radio sources is only *additive*, not *subtractive*; in other words, unlike sound, with light

---

[40] Herbert Dingle, "Relativity and Electromagnetism: An Epistemological Appraisal", Philosophy of Science, 27, (1960), p233-253.
http://www.jstor.org/stable/185967
[41] Thomas Young, "The Bakerian Lecture: On the Theory of Light and Colours", Philosophical Transactions of the Royal Society of London (1776-1886),Volume 92, (1802), pp12-48.
http://books.google.com/books?id=-XAXAQAAMAAJ&pg=PA140

interference, two signals can only add together, not subtract from each other. I only know of two publications that claim light interference for radio, one is an 1888 paper[42] of Heinrich Hertz, and the other is a 1897 paper[43] by Augusto Righi, for which there is no English translation that I am aware of. Hertz's claim is hard to visualize without video. Hertz finds that at certain points the signal from two sparks cause the two balls on a secondary conductor to have the same electric potential and so they do not spark, while everywhere else, they have a different potential and so do spark. But this fits with a corpuscular "additive-only" theory for light, since the light particles are still there, but simply in the same quantity in each ball. It's not an absence of signal; it's an absence of spark. Righi claims, in his work, that waves of radio light interfere just like visible light in equivalent optical experiments using thin films and mirrors, which to me implies a reflection phenomenon; a more convincing experiment would not require any reflecting apparatus, but would simply show how two light sources create subtractive interference just like two sound sources do. That Righi's work has not yet, as far as I know, been translated to English, and that no "radio interference" experiments have ever been reported by English speaking authors, shows a clear lack of effort to prove this point to the public.

---

[42] H. Hertz, "Ueber die Ausbreitungsgeschwindigkeit der electrodynamischen Wirkungen", Annalen der Physik, Volume 270 Issue 7, p551-569.
http://books.google.com/books?id=_D0bAAAAYAAJ&pg=PA551
English translation:
Heinrich Hertz, tr: D. E. Jones, "On the Finite Velocity of Electrodynamic Actions", "Electric Waves", 1893, 1962, p59.
http://books.google.com/books?id=EJdAAAAAIAAJ&pg=PA59
[43] Augusto Righi, "L'ottica delle oscillazioni elettriche", Bologna, 1897.
http://books.google.com/books?id=QRU6AQAAIAAJ
English translation: "The Optics of electrical oscillations".

Polarization

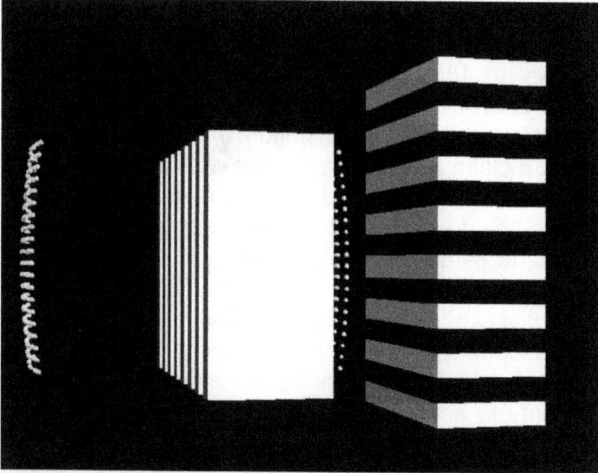

Figure 3.8. Image from a simulation showing light particles filtered by a vertical and then horizontal polarizer.

Polarization is a similar reflecting phenomenon. Light entering polarizing material has its *direction* restricted (not, as is currently claimed by many authoritative sources, that rays of light all "vibrate" in the same direction). Polarizing material can easily be thought of as being made of long columns of planes (see figure 3.8). When the polarizer is held vertically (the columns are vertical), any light rays with some sideways direction (usually referred to as the "X" dimension) are reflected and/or absorbed by the polarizing material, but the entire 180 degrees of vertical rays (with a "Y" dimension component) pass through unreflected. The opposite is true for a polarizer held horizontally; all rays with vertical directions are reflected back or absorbed by the polarizer walls, while rays in any of the 180 degrees of horizontal direction pass through. By putting two polarizers together at 90 degrees to each other, almost all ray directions are filtered out by reflection and absorption, except those passing directly through with no X or Y component to their direction (only a "Z" or forward component).

Refraction

Refraction is probably the result of light particles that are reflected by other light particles (within atoms) in a denser material. When this happens, the direction of the light particles changes; if Z is the direction perpendicular to the refracting material, the X and Y components of the rays are made less relative to the Z. This is because of collisions the particles of light experience. Rays perpendicular to the material do not change angle because they can only reflect directly back (there is no X or Y direction component to lose from collisions).

Double refraction

Double refraction may be the result of light particles reflecting off of two perpendicular atomic planes that are tilted slightly in the X and Y planes so that light reflects off them and then continues through in the Z direction, but with a different X and Y angle. One set of rays is polarized relative to the Y and the other relative to the X direction, so two double refracting crystals can act like a polarizer in blocking most of the light with X and Y components. A good experiment is to beam a low-cost last down onto a tilted glass slide, and see that two laser point images are made: one that passes through the slide, and a second that is reflected by the slide. You can see how rotating the slide causes the reflected image to rotate.

I may be wrong on one or more of these explanations, but who can doubt that there is not a perfectly fine particle reflection explanation for refraction or any other phenomenon of light? Certainly D2B owners and probably D2B consumers already know all this and much, much more, but aren't telling the poor D2B excluded anything.

The neutron is probably a hydrogen atom

Just like the bizarre D2B rule that nobody can reveal how all matter is made of light particles, so it may be that another minor misleading claim is that the neutron is not just a hydrogen atom. James Chadwick, the person who first named the neutron supposed the neutron to "consist of a proton and an electron in close combination" with a mass "slightly less than the mass of the hydrogen atom"[44,45] If neutrons are actually Hydrogen atoms, then so-called neutrons could be combusted with Oxygen just like Hydrogen atoms. Certainly a neutron is made of light particles, and perhaps the emission and absorption spectrum are the same as that of Hydrogen.

Why no telling the public about the details of evolution, the history of science, and the future?

Where is the movie "Evolution" for the large screen? How about the movie "Science"? And then the most interesting movie of all: "The Future" – where are they? As is the case I make above, the silence is one of the best pieces of evidence of some kind of large scale conspiracy of silence - of creating a Pol-Pot kind of society where a group of elites sees and knows everything, while purposely leaving the public in absolute ignorance and misleading them with obviously false "official" stories and explanations.

I have spent the last 8 years researching the history of science. Making my "Universe, Life, Science, Future" database and videos has been a full time hobby that I have worked consistently on for all that time. So it gave me an unusually good perspective on the course of science history as experienced and recorded by many of the great scientists of the past.

---

[44] J. Chadwick, "Possible Existence of a Neutron", Nature, vol 129, 1932, p312.
http://www.nature.com/nature/journal/v129/n3252/pdf/129312a0.pdf
[45] J. Chadwick, "The Existence of a Neutron", Proceedings of the Royal Society of London. Series A, Containing Papers of a Mathematical and Physical Character, Vol. 136, No. 830 (Jun. 1, 1932), pp. 692-708.
http://www.jstor.org/stable/95816

You owe it to yourself to see my excluded version of the complete story of the universe, the evolution of life, the history of science, and a projection into the far future all in a 10 minute free movie at ulsfmovie.org. The other progressively longer versions have far more details and important information, also for free.

"Non-Euclidean" surface geometry unlikely to apply to the universe

The origins of so-called non-Euclidean geometry start with the Russian mathematician Nikolay Lobachevsky, who, in 1829, was the first to publish a non-Euclidean geometry[46]. The goal of "non-Euclidean" geometry is to disprove one or more of Euclid's postulates, and what grew out of this effort was the rise in the popularity of the mathematical field of *curved surface geometry*.

Lobachevsky published the first known instance of the famous claim that a triangle made of curved lines may have angles that add to more than 180 degrees (for example on the surface of a sphere), an apparent violation of the Euclidian postulate that all angles of a triangle must add to 180 degrees (or pi radians). Another case is when the angles of a triangle add to less than 180 degrees (for example on the surface of a hyperboloid). One problem with this is that there is not a single angle formed between two curved lines (even on a plane); the angle changes the closer to the intersection one measures.

This new idea of restricting space to a curved surface was then applied to space in the universe. The argument was that space in the universe may be "curved", but like looking at a small part of a very large curve, only appears to be straight. There were early critics of this new and unlikely idea that space in the universe might be somehow "curved",

[46] NI Lobachevsky, "On the foundations of Geometry", Kazan Messenger, 1829.

including the very influential Hermann Helmholtz[47], which I hope to describe in more detail in the future.

Even if we accept that there are some shapes and/or spaces that are not described by, or that violate one or more of Euclid's postulates, and so can be called "non-Euclidean", there is a simple truth that any "non-Euclidean" surface geometry is only a subset of unrestricted space. What is being called "non-Euclidean" geometry is more accurately called "curved surface geometry", because all that this geometry does is to restrict the possible values of the variables that represent position (traditionally for 4 dimensions the variables x, y, z, and t, or for two dimensions u and v). The possible positions are often limited to the surface of some geometrical shape like a sphere, or hyperboloid. So space in a surface geometry is a *subset* of unrestricted space, and I think that the traditional unrestricted three dimensional space is the most accurate model of space for the universe.

To D2BW excluded ears, it may sound overly conservative, but in my opinion (and no doubt in the opinion of those receivers of D2BW who control many micro and nano scale particle devices), the traditional view of space as described by three variables (x,y,z) is not only more accurate and logical than restricting space to a curved surface geometry, but is far more simple.

One theory I am putting forward here is that this new abstract curved-surface geometry was and still is embraced and funded by many D2BW owners and consumers (who don't make any use of it

[47] Helmholtz, H., "Über die tatsächlichen Grundlagen der Geometrie", Verhandlungen des naturhistorisch-medicinischen Vereins zu Heidelberg, 4, 1866, pp. 197-202.
http://books.google.com/books?id=hksDAAAAYAAJ&pg=PA197
English: "On the Actual Foundations of Geometry"
Some parts translated in:
Joan L. Richards, "The Evolution of Empiricism: Hermann von Helmholtz and the Foundations of Geometry", Brit. J. Phil. Sci. a8 (1977), p235-253.
http://www.jstor.org/stable/686808

themselves) because it helps to distract and keep excluded people away from understanding and being interested in science.

Special and General Theory of Relativity very unlikely

This is not to say that I reject the claim that nothing can move faster than the speed of light, or that gravitational and inertial acceleration are not equivalent, but just that I reject 1) that light is massless, 2) that the speed of light is always constant, 3) that mass and motion can be changed into each other as implied by equations of momentum and energy, 4) the idea that space and or time contracts or dilates depending on the motion of matter, and 5) that a surface geometry applies to space and time in the universe.

I reject the theory that light is massless in favor of light being made of material particles which are the basis of all matter (as the smallest known particle). This is obvious when we see light emitting from a burning candle and the candle made smaller in mass as a result. So I define the label "photon", not as a quantum of energy, but as a material light particle which is the basis of all matter.

I think that light particles can reflect off of each other, and the evidence of this is simply that light reflects off of objects. So for light to reflect, it seems likely that the velocity of light (which is simply space covered over time) can not only be changed, but can be 0. We can imagine a particle of light reflecting off of other particles in a space that becomes filled with more and more light particles. As the number of light particles in the space increases, the space and time between collisions decreases, and so eventually, a light particle may be held motionless, packed together with many other light particles. In fact, we may be sitting on an atom factory. Atoms may form near the surface of planets and stars where space becomes less dense and there is room for atoms to form without being torn apart by collision. In 1940,

Haxby et al at Westinghouse showed that even light particles can cause atomic fission[48]. So like molecules, atoms too, may be objects that can only exist at relatively low temperature and density.

Many people hold up the chain-reaction of uranium fission, and the equation $E=mc^2$ as proof of the Theories of Relativity and the equivalence of energy and matter, but atomic fission was the result of experiments done by Enrico Fermi, Otto Hahn and Lise Meitner, not the result of trying to confirm a physics theory like relativity. In addition chain-reactions can be just as easily explained with a particle collision chain-reaction theory, and as I explained above, any equation of energy or momentum presumes that mass and velocity can be exchanged. While this matter and motion interchangeability may be a useful generalization to describe some physical phenomena, it seems, in my mind, unlikely that motion can spontaneously change into matter, or matter spontaneously change into motion; I find more likely the theory that motion is simply *transferred* between pieces of matter by collision. It may seem hard to believe that a small motion can lead to an apparently much larger motion, but typical combustion (burning of hydrocarbon and oxygen) involves the same exact phenomenon- a small motion in for a large and sometimes long duration motion out. One example is making an opening in a container with a material under high pressure; a little motion in results in much more motion out. In this view, light particles are trapped inside some high pressure material container within atoms, and when this container is broken, the light particles fly out. Another theory is that light particles have a lot of motion within atoms, but that this motion is simply directed in a closed orbit within the atom, and chain reactions simply knock them out of orbit so they fly out in all

---

[48] R. O. Haxby, W. E. Shoupp, W. E. Stephens, and W. H. Wells, "Photo-Fission of Uranium and Thorium, Phys. Rev. 58, 92-92 (1940). http://prola.aps.org/abstract/PR/v58/i1/p92_1

directions. Either way, I think that all chain reactions like atomic fission and combustion involve the same simple process of atoms separating into their source light particles. Particle collisions inside an object at our scale may give light particles within them a much higher velocity than the surrounding matter, and the duration of the collision may be so short that the added motion of the object at our scale in that time is too small to measure. One big question that modern people do not often address is where all those light particles come from. The current view simply waves them away as "energy", or has light particles emitted from combustion as being lost from electrons, but I think it seems likely that entire atoms (including protons and "neutrons") may be separated in to their source light particles in combustion. If true perhaps some apparently non-atomic exothermic (or "exophotonic") reactions might result in radioactive atoms, different atoms, or different atomic isotope products. Probably D2B owners already know the answers to these questions, but like D2BW, are not sharing them with the public.

In addition I reject the Theories of Relativity because I reject the theory of space and time dilation and contraction, and I reject the theory that time depends on the motion of matter. For example, a clock may move more slowly under water, but that doesn't mean that time in the universe is slower, or that time as experienced by the clock is slower. I think that there is just one time for the entire universe. It seems logical to me that if the time is 10 o'clock here, it is also 10 o'clock in every other galaxy. Herbert Dingle identified a simple problem with the so-called "twin paradox" claim of time dilation in which one twin ages more slowly because they are moving faster than the other twin: how could one twin be moving at a different velocity compared to the other twin when their motion is relative to each other?[49]

---

[49] Herbert Dingle, "Relativity and Space Travel", Nature, **177**, 782 (1956).

There are a number of reasons why the General Theory of Relativity is popular. I think that the number one reason is that those who own and control remote neuron writing, as crazy as it sounds, want to keep the public as far away from understanding the details of seeing, hearing and sending thought images and sounds as possible-they simply don't want any competition, or for the public to see all the past images of many unpunished violent crimes. Secondly, very few people have ever learned calculus or how math is used to determine future locations of pieces of matter. Perhaps another reason for the popularity is that many people think that a new theory is an improved theory, but just because a theory is new doesn't mean that it is a better theory. There are plenty of examples where the new theory was certainly no better and many times much worse than the earlier most popular theory: the rise of Christianity and Islam in the Mediterranean replacing Polytheism around 400 and 700 AD respectively, the rise of Nazism and fascism in parts of Europe during the 1930s, etc. In addition, Einstein was a colorful character, and represented, to many people, the opposition to the rising Nazi movement in Europe. Perhaps Einstein's theory winning, represented the anti-Nazis winning in the minds of many people, and so they agreed to support these unlikely theories. Many Nazi's famously rejected Einstein's theory of Relativity and so opposition to Relativity, became associated with support for the Nazis, even though much of the opposition to theories like relativity are from a desire for truth and accuracy, and from a fear of inaccurate claims becoming popular.

There have been many famous critics of Einstein's theories of relativity, some memorable critics are Charles Lane Poor[50], William Pickering[51], and Herbert Dingle[52].

---

[50] Charles Lane Poor, "Gravitation Versus Relativity", Putnam, 1922. http://archive.org/details/gravitationversu00poorrich
[51] Pickering, W. H., "Shall we Accept Relativity", Popular Astronomy,

One related and relevant question of these centuries is: "Did the person receive direct-to-brain windows or were they one of the excluded?". For example, did Isaac Newton get D2B? Did Beethoven? Did Ben Franklin? Did Abe Lincoln? Did Einstein? Did JFK? Did Marilyn Monroe? Did Elvis? Either way, the truth about D2B significantly changes what we know from the public story. They either watched videos in their eyes, or knew nothing about it, and were the victimized excluded. Did Einstein get D2B? It seems likely to me that he did, and compared to relativity, that theory is probably the more accurate one.

One of the saddest parts about the rise of non-Euclidean geometry and the theory of relativity is that it has been one of the best and most successful efforts against arousing public interest in science in years. Average people are shushed away from the big curtain of science and technology with the excuse that they are not qualified to understand the universe, science and technology, when the exact opposite is true. Most of the universe is very simple and logical in my opinion. For example, that all matter is made of light particles – how much more simple could it be? That our fate is to build a globular cluster, that thought images and sounds can be seen, heard and sent, etc. This reign of non-Euclidean geometry and relativity has been a century of dogma, but is small when compared to the millennia of religious dogma.

In my view, the much more likely theory is the "billiard-ball" universe, where light particles just move around the universe colliding into each other. Even the phenomenon of gravity, as we experience it, being pushed down to the surface of the Earth, may be the result of many particles colliding with us,

Vol. 30, 04/1922, p.199.
http://articles.adsabs.harvard.edu/cgi-bin/nph-iarticle_query?bibcode=1922PA.....30..199P
[52] Herbert Dingle, "Science at the Crossroads", Martin, Brian and O'Keefe Ltd, 1972.

particularly from above us. If gravity is just the result of particle collisions in a dense matter field, then any "anti-gravity" would be doubtful. The only way to move away from a dense matter field would still be the only method we know; to use material particle collisions to push our way out. It's impossible to sum up the positions and motions of millions of particles, but overall large-scale motions can be generalized using the simple equation for the force (or collective effect) of gravity. Simply using "iteration" (modeling a universe of many particles and measuring their mutual gravitational influence on each other for each frame of time) using Newton's simple $A_2 = Gm_1/r_2$ equation, and including collisions, is probably the most simple, fastest, and accurate method to model the universe. Almost certainly that, and the simple F=ma inertial motion law are what the neuron owners have used for centuries when, for example, trying to determine how contracting a muscle will cause a body to move in the future.

The truth is that history is filled with millions of "mistaken beliefs" and just simply "lies". There is not space to list them all, but some are: the claims that the Sun goes around the Earth, of Angels, of a Heaven, of Demons, ghosts, spirits, witches, that Jesus rose from the dead, Moses parted the Sea, Lee Oswald killed JFK, Islamic hijackers brought down the World Trade Centers, that people didn't figure out how to see and hear thoughts, and we can add to the garbage bin of mistaken beliefs: the Big Bang expanding universe theory, background radiation, and both the special and general theory of relativity.

So what is the most accurate interpretation of the universe in my view?

With so many lies, frauds and mistaken beliefs, what is the average person to believe are the most accurate theories? First, I am sure, that those people who own D2B and many D2B consumers must have a much better understanding of the universe than the public is being shown. For myself, as I stated above, I support an "all inertial" "billiard ball" universe, where matter and motion are never created or destroyed, and are always conserved, but I think clearly that Newton's law of gravitation is the best and most useful equation to use when modeling matter in the universe. But even Newton's simple law may not help us to predict what living objects in the universe may do in the future. As I say above, clearly the Theories of Relativity, of time and space contraction and dilation, the expanding universe, the electromagnetic wave theory of light, and most of the claims of psychology and religions are all false. Instead, all matter being made of light, which is a material particle is the most accurate truth in my opinion. Beyond that, the theory of a single common ancestor for all of life on Earth and of natural selection (the Theory of Evolution) is one of the few theories that is accurate and will survive the truth about RNRAW. One theory that I find impressive is that, since the universe is probably infinite is time and size, perhaps at the scale of galactic clusters, or at the atomic scale, there are collective objects that we would compare with ourselves as living objects.

I'm only interested in the truth and telling everybody else the truth, I'm not interested in lying to and tricking people. Another point is that I am not looking to ridicule those who believe religions and other inaccurate theories, but instead trying to win them over to the more accurate truth. I feel sorry for many of those people, because they have been tricked, lied to, duped, fooled with the lies and unlikely claims of religions, and by deliberately dishonest scientific claims such as the expanding universe theory, that light is not the most basic atom, that people haven't figured out remote neuron reading and writing long ago in the past, etc.

Chapter 4
Timeline: A 700 Year Secret?

Here is a timeline that presents some evidence of secret remote neuron reading and writing- a secret that may go back to even 1300, as crazy as that sounds.

1038 CE
The pin-hole camera (or camera obscura) is first described by Alhazen (Ibn al-Haytham)[53].

1208 CE

Figure 4.1. Robert Grosseteste claimed that all matter is made of light in 1208.

Robert Grosseteste (fig. 4.1) writes that all matter is made of light.[54] This is around the time of the first Universities (Paris, Oxford, etc.) in Europe, and is evidence of a collective exploration of science which, for smart people, would quickly focus on light particles, communication, electricity, micromachining, motors, cameras, and neurons.

[53] "Ibn Al-Haytham, Abū ʿAlī Al-Ḥasan Ibn Al-Ḥasan." *Complete Dictionary of Scientific Biography*. Vol. 6. Detroit: Charles Scribner's Sons, 2008. 189-210. *Gale Virtual Reference Library*. Web. 11 Sep. 2012.

[54] Robert Grosseteste, "De Luce", 1208.

1589

Figure 4.2. William Byrd wrote "your minde is light" in 1589.

William Byrd (fig. 4.2) publishes "Songs of Sundrie Natures"[55], which is the earliest evidence I have yet found for knowledge of neuron reading. Byrd writes: "Your minde is light", "the Sunne with his beames", and "we were out and he was in". Byrd ends this very interesting song with "Two dayes before it was begoonne" which may imply that neuron reading or writing was first realized, two centuries before 1589 - which would be 1389. Clearly, also here, there must be some prohibition against saying that all matter is made of light by this time under some more powerful authority. It could be the owners of the Church, but more likely it is the owners of the Government, Military, and Communications. This song also contains the word "gravitie" 100 years before Newton.

All I can say is that it becomes somewhat easier over time to recognize the language of those who receive direct-to-brain windows- even though we excluded don't receive regular, consensual, D2BW. Year 1500 or even year 1200 for that matter seems like a ridiculously long time ago, and are viewed as a

---

[55] W. Byrd, "Songs of Sundrie Natures", 1589.

stone age to most people - but knowing that remote neuron reading and writing is easily 200 years old - you can see that wealthy people who developed this technology could have simply kept it to themselves - and viewed poor and middle-income people (those who must work for money) with very little concern. Now, in the 2000s, clearly many poor people receive direct-to-brain windows, but the poor may have been excluded for centuries. Most poor, middle and even some wealthy humans still are completely unaware of RNRAW and may fill some sadistic, abusive, and highly sexually pleasing entertainment purpose for those who distribute and receive direct-to-brain windows.

1600

Figure 4.3. Giordano Bruno, victim of the war on truth and science.

Giordano Bruno (fig. 4.3) is murdered, burned at the stake, in part, for his view of a moving Earth. This murder of scientists, engineers, teachers, those who tell the truth, and non-religious people, has been the theme on Earth for millennia and continues into the 2000s, but now the murdering and torturing is mostly done remotely with microscopic particle devices.

1664

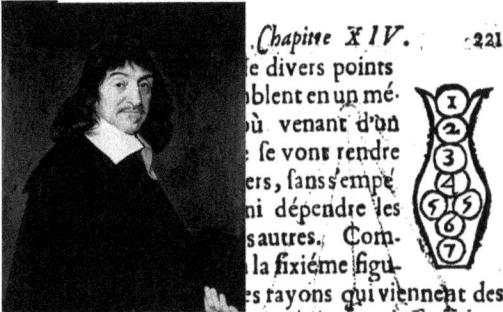

Figure 4.4. Image of René Descartes and a figure from "Le Monde".

René Descartes' "Le Monde" (fig. 4.4) identifies the two major theories for light, the wave and corpuscular theory[56].

1672

Figure 4.6. Isaac Newton, and figure from his first letter to the Royal Society.

[56] Descartes, R. Le Monde ... Ou Le Traité De La Lumière Et Des Autres Objets Principaux Des Sens, Avec Un Discours De L'action Des Corps Et Un Autre Des Fièvres, Composez Selon Les Principes Du Même Auteur. Michel Bobin et Nic. le Gras, 1664.
http://books.google.com/books?id=DHEPAAAAQAAJ
English translation: Rene Descartes, Translated by Michael S. Mahoney, "The World or Treatise on Light", Chapters 13 and 14.
http://www.princeton.edu/~hos/mike/texts/descartes/world/worldfr.htm

Isaac Newton (fig. 4.6) more clearly and firmly establishes a "corpuscular" theory for light.[57] The most obvious view that light is made of particles will collapse in the beginning of the 1800s (because of the rise in popularity of the transverse wave theory for light of Thomas Young). But a particle theory for light will be partially revived for a third time by Planck and Einstein in the beginning of the 1900s. Apparently it may be that when the century turns, every one hundred years, at least one, significant science contribution is released to the public by the D2BW owners. Certainly the D2BW owners play a large part of the suppression of the simple "light is made of particles" truth to maintain their ridiculous 700-year head-start and monopoly of light particle technology.

1678

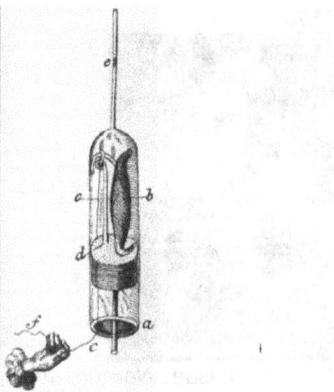

Figure 4.5. Image of Jan Swammerdam and drawing of his 1678 work.

The earliest publicly known direct neuron writing: Jan Swammerdam contracts a frog leg muscle using

---

[57] Isaac Newton, "Draft of 'A Theory Concerning Light and Colors'", Feb 6, 1671/2, in English, c. 5,137 words, 14pp. Shelfmark: MS Add. 3970.3, ff.460-466 Location: Cambridge University Library, Cambridge, UK
http://www.newtonproject.sussex.ac.uk/view/texts/normalized/NATP00003

two different metals (fig. 4.5).[58] Most people are not aware of how far back into the past neuron writing actually goes. It simply was not called "neuron writing".

It may be that the exponentially growing, well organized government, military, communications and university secret research into remote neuron reading and writing (none of which gets published) is highly developed by this time (the late 1600s). If true, then almost all science here and later: Galvani, Joe Henry, Heinrich Hertz, etc. is actually probably 1) neuron consumers publishing ("leaking") old findings, or 2) excluded re-inventing secret findings of the past.

1689

Although, very minor, a hint that there were "insiders" and "outsiders" at this time can be found in a 1980 book "Norton Anthology of Western Music" which states of Henry Purcell's opera "Dido and Aeneas": "...It was first performed in 1689 by the pupils at a girl's boarding school in Chelsea, a suburb of London, with a few outsiders probably pressed into service for the men's parts. ...".[59] Notice the possible sexual double meaning of "men's parts".

[58] John Joseph Fahie, "A History of Electric Telegraphy, to the Year 1837", E. & F. N. Spon, 1884.
http://books.google.com/books?id=0Mo3AAAAMAAJ

[59] Claude V. Palisca, "Norton Anthology of Western Music", fourth edition, 2001, p419.

1772

Figure 4.7. Joseph Priestley, published a history of Opticks in 1772 that is in some ways more advanced than modern science, because, unlike the majority view today, Priestley viewed light as being made of material particles.

Initially I thought that seeing and hearing thought probably only went back to the 1950s, but then earlier: to 1910. That seemed at first shockingly early to me. Then I found more evidence, and realized that RNRAW and D2BW may go back to at least 1810. But then Joseph Priestley (fig. 4.7) in his "History of Opticks"[60] uses the key word "tenable" and that was 1772.

---

[60] Joseph Priestley, "The history and present state of discoveries relating to vision, light, and colours.", Leeds: n.p., 1772.
http://echo.mpiwg-berlin.mpg.de/ECHOdocuViewfull?mode=imagepath&url=/mpiwg/online/permanent/library/BVC1P0A1/pageimg&viewMode=images

1791

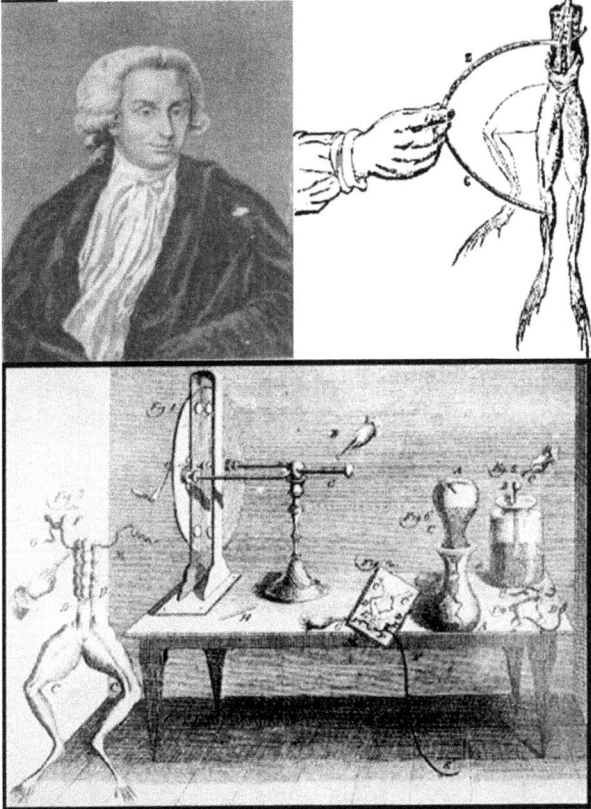

Figure 4.8. Portrait of Galvani and images from his 1791 paper.

Remote neuron writing[61]: Luigi Galvani makes a frog leg muscle move by touching the frog nerve with a scalpel while an assistant cranks a remote spark generator (fig. 4.8). Light from the spark reaches the scalpel, takes the form of electricity (photoelectric effect) and causes the frog leg to contract. This is the first example of using light particles to contract a muscle remotely- all the way

---

[61] Luigi Galvani, Elizabeth Licht, Robert Green, "Commentary on the Effect of Electricity on Muscular Motion", Waverly Press, 1953.

back in 1791. How many people know that remote neuron writing goes back at least to 1791? Isn't that something very important that we should know? Galvani also put the frog leg in between one piece of copper and one piece of tin to make the muscle contract. This will lead to the first electric battery. This 1791 publication is a clear indication that remote neuron activation may be widely known, although secretly, by this and all later times. The word "galvanized", taken from Galvani's last name, has come to have multiple meanings, for example "to be strongly set in opinion", but one secret meaning is that when somebody is murdered using remote neuron writing (for example to contract their lung muscles, or their heart) people say that they were "galvanized".

So you can say without any hesitation to anybody that "remote neuron writing goes back at least to 1791", because that is a clear and published fact.

1800

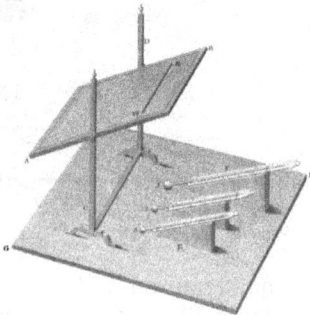

Figure 4.9. Portrait of Herschel and image from Herschel's 1800 paper (note the three thermometers).

Invisible light: William Herschel (fig. 4.9) reports that an invisible part of the spectrum (the infrared) heats a thermometer more than any other part of the spectrum.[62]

---

[62] William Herschel, "Investigation of the Powers of the Prismatic Colours to Heat and Illuminate Objects; With Remarks, That Prove the

1801

The absolute length and frequency of each vibration is expressed in the table; supposing light to travel in $8\frac{1}{2}$ minutes 500,000,000000 feet.

Colours.	Length of an Undulation in parts of an Inch, in Air.	Number of Undulations in an Inch.	Number of Undulations in a Second.
Extreme -	.0000266	37640	463 millions of millions
Red - -	.0000256	39180	482
Intermediate	.0000246	40720	501
Orange - -	.0000240	41610	512
Intermediate	.0000235	42510	523
Yellow -	.0000227	44000	542
Intermediate	.0000219	45600	561 ($= 2^{47}$ nearly)
Green - -	.0000211	47460	584
Intermediate	.0000203	49320	607
Blue - -	.0000196	51110	629
Intermediate	.0000189	52910	650
Indigo - -	.0000185	54070	665
Intermediate	.0000181	55240	680
Violet - -	.0000174	57490	707
Extreme - -	.0000167	59750	735

Scholium. It was not till I had satisfied myself respecting all these phenomena, that I found in Hooke's Micrographia, a pas-

Figure 4.10. Thomas Young and his table showing the wavelengths (particle spacings) and frequencies for various colors of light.

Thomas Young (fig. 4.10) shows that color is related to the frequency of light, and measures the wavelengths (particle spacings) for different colors.[63] In determining frequencies for light Young made a valuable contribution to human progress, but his opposition to a particle theory for light helped to set back the human species for more than a hundred years.

Different Refrangibility of Radiant Heat. To Which is Added, an Inquiry into the Method of Viewing the Sun Advantageously, with Telescopes of Large Apertures and High Magnifying Powers.", *Philosophical Transactions of the Royal Society of London* , Vol. 90, (1800), pp. 255-283.
http://books.google.com/books?id=dlFFAAAAcAAJ&pg=PA255
[63] Thomas Young, "The Bakerian Lecture: On the Theory of Light and Colours", Philosophical Transactions of the Royal Society of London (1776-1886),Volume 92, (1802), pp12-48.
http://journals.royalsociety.org/content/q3r7063hh2281211/?p=422e575 bae414c9a974a16d595c628d0Ï€=24

1810

Figure 4.11. William Hyde Wollaston.

William Hyde Wollaston (fig. 4.11), and the Royal Society, probably definitely were involved in remote neuron reading and writing technology. The Royal Society has done more than most other societies to educate and inform the public about secret scientific advances over the course of history. Around this time Wollaston gives a lecture in which he uses keywords like "thought", "render", and draws attention to vibrations that can cause sounds in the ear.[64]

The famous story "The Strange Case of Dr. Jekyll and Mr. Hyde" of 1886 may have been an effort to point blame at "Hyde" Wollaston, the Royal Society, or science in general for the countless remote particle abuses, or to hint at the involvement of the Royal Society with the development and use of neuron reading and writing technology.

---

[64] William Hyde Wollaston, "The Croonian Lecture", Philosophical Transactions of the Royal Society of London , Vol. 100, (1810), pp. 1-15 http://books.google.com/books?id=2xJGAAAAMAAJ

1816

Figure 4.12. Joseph Niépce and the first public photograph.

First photograph: Joseph Niépce makes the first publicly known photograph. The above image (fig. 4.12) on the right is from 1826 by Niépce (on the left) and is the first publicly known permanent photograph. But clearly the microscopic wireless floating and flying cameras must have been developed very early – perhaps even by 1200 as absurd as that sounds.

1827

Figure 4.13. André-Marie Ampère.

In the famous paper by André-Marie Ampère (fig. 4.13), which describes how two current carrying wires move toward or away from each other depending on the direction of the current, one paragraph contains a sentence using the words "suggère" (suggest) and "contractions musculaires" (muscle contractions).[65] The idea of using electricity to contract muscles also may imply the secret research of artificial muscles and walking robots that move much like humans by this time. An artificial muscle could easily be constructed using this simple phenomenon by drawing wires together to contract a flexible rubber material.

Félix Savary describes the phenomenon of electrical oscillation[66]. How an electric current oscillates between a capacitor (a Leyden jar) and an inductor. This is a basic part of wireless communication.

1842

Figure 4.14. Joseph Henry.

[65] André-Marie Ampère, "Théorie des phénomènes électro-dynamiques, uniquement déduite de l'expérience. ", Méquignon-Marvis, 1826
http://www.ampere.cnrs.fr/ice/ice_math.php?typebookDes=Oeuvres&bdd=ampere&bookId=23
http://gallica.bnf.fr/ark:/12148/bpt6k29046v
A partial English translation is in:
Tricker, R. A. R., "Early Electrodynamics - The First Law of Circulation", (Pergamon, NY), 1965, p155-200.
[66] Félix Savary, "Mémoire sur l'alimentation", Annales de Chimie et de Physique, 1827, 34:54-56.
http://books.google.com/books?id=QaQwAAAAYAAJ&pg=PA54

Joseph Henry (fig. 4.14) describes the basis of radio when he reports magnetizing a needle that is "...7 or 8 miles away..." by electrical induction from an electric spark.[67]

1846

Figure 4.15. John Thomas Perceval.

John Thomas Perceval (fig. 4.15), son of murdered Prime Minister of England Spencer Perceval, was apparently excluded, had some kind of outburst, and was held in 2 psychiatric buildings for about 4 years. After being released Perceval published two books (1838, 1840), and spent his life working to grant people locked in hospitals better protections. Clearly Perceval hints and uses double-meaning words to promote the "end the neuron lie" platform.

For example:

On 11/27/1846 Perceval writes in the Visitors' Book of Bethlem Hospital:

"Amongst the most painful of these circumstances was the constant sight of heavy bars to my window, ...I think the Committee might safely remove these

---

[67] Joseph Henry, "On Induction from Ordinary Electricity; and on the Oscillatory Discharge.", Proceedings of the American Philosophical Society, vol. II, 1842, p193-196.
http://books.google.com/books?id=5AIwAAAAIAAJ

bars, and substitute windows with small sashes in iron frames-or adopt in some cases, the plan pursued in many private asylums, of having Venetian blinds to the windows. ...". Note the double meaning of "bars to my window" - like something barring the way to receiving direct-to-brain windows - even then in 1846 they were called "windows" - long before Windows 3.1 or X-Windows. Note also "blinds to the windows" - those who don't get D2B are many times referred to as the "blind". "...I consider that society or the Legislature, who shut up patients not only for their own benefit ... but for the benefit of society as well . . . in a manner are compelled, in doing so, to violate the liberty of the subject...". We all recognize "shut up" in modern times from the contemporary Nazistic-Bush-Cheney movement - most clearly demonstrated by sources like Fox News. Here, notice again, "shut up" has multiple meanings - being locked up, but also that people are made to "shut up" about talking about direct-to-brain windows and the neuron secret- then at the ripe old age of perhaps 500 years. It's obvious that the owners (and many consumers) of the neuron technology want to preserve their monopoly and advantage of getting and sending thought images, sounds, and information in addition to remote muscle moving. Ironically, there is no possibility of shutting up *your* information circulating around *their* eyes- but plenty of chance of *their* info being shut up from *your* eyes, in addition to no shutting up of all their constant remote particle murders, assaults, molestations and lies. Also, the idea that the psych establishment is used to shut up or lessen the popularity of people with views contrary to those in power is a theme for much of the history of the D2B secret.

Other hints are minor: "That he knew all my thoughts..." and "...my necessities, were not once consulted, I may say, thought of." - note "I may say" - like he does or does not have permission to reveal some information. Later he writes "I cannot say ..."[68].

1861
October 26

Figure 4.16. Philip Reiss

Philip Reiss (fig. 4.16) goes public with the first known microphone, telephone and speaker. Sound can now be converted to electricity and back to sound again.[69] Quietly sending sound over longer distance is now possible. Reiss dies at a young age and may have been remotely murdered; made an example of, symbolically, for going public with encoding and decoding sound in electricity (the telephone).

---

[68] Szasz, Thomas (editor) (1975 (1973)). "The Age of Madness: A History of Involuntary Mental Hospitalization Presented in Selected Texts. ", London: Routledge & Kegan Paul Ltd. ISBN 0710079931.
[69] Silvanus Phillips Thompson, "Philipp Reis: inventor of the telephone: A biographical sketch, with ...", 1883.
http://books.google.com/books?id=7uQOAAAAYAAJ

1864 CE

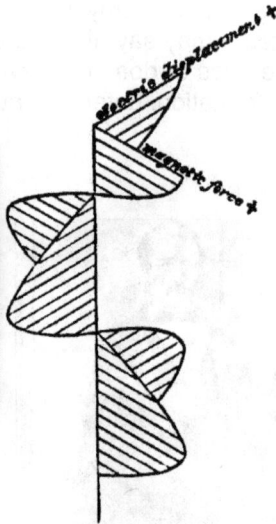

Fig. 66.

Figure 4.17. James Clerk Maxwell and a drawing of an "Electromagnetic" light wave from his book of 1881[70].

James Clerk Maxwell publishes his Electromagnetic theory of light (fig. 4.17)[71], which is, with all due respect, a very unlikely theory, where light is made of two sine waves: an electric wave, and a magnetic wave, which are at 90 degrees to each other, in an aether medium. Beyond the Michelson experiment of 1881, which is evidence against an aether medium for light, this theory is easily disproven by the fact that light with low frequencies, like radio, can be focused by a concave object (for example a mirror) to a point. It seems doubtful that light in the radio with a 10 meter or more wave length could be focused to a point of

[70] James Clerk Maxwell, "A treatise on electricity and magnetism.",vol. 2, 1st ed, Oxford, 1881, p390.
http://books.google.com/books?id=gmQSAAAAIAAJ&pg=PA390
[71] James Clerk Maxwell, "A Dynamical Theory of the Electromagnetic Field", Royal Society Transactions, Vol. 155, 1865, p. 459-512.
http://books.google.com/books?id=xVNFAAAAcAAJ&pg=PA459

variable distance unless radio light has no amplitude. Beyond that, a straight line, material, particle interpretation for light is far less complicated and still can explain all known phenomena of light. The explanation of radio as light particles emitted from particle collisions in electric current seems an obvious but mysteriously missing alternative theory. Light is emitted from rubbing two objects together and from exothermic chemical reactions, but light is not called an "electromagnetic", "mechanical", and "chemical" wave. Light is better described as being made of material particles, and the so-called "electromagnetic spectrum" as the "spectrum of light particle frequencies".

Shockingly, this electromagnetic theory of light is still the most popular and "official" view 148 years later. This work begins the "trick the excluded with math" years. But this at least reawakens the idea that the public should gain access to invisible frequency particle communication (radio).

1869 CE

Early evidence of neuron reading and writing can be seen in the 01/30/1869 letter titled "Brain Waves: A theory" (fig. 4.18) by one "J.T.K." to the Editor of "The Spectator"[72]. This paper contains numerous word plays and hints about how remote neuron reading and writing is used to trick women into sex, trick men into doing violence, to make people do unusual activities that make them appear "insane", how letters are put together to spell words, and many other hints. Note that this is before Edison in 1885 and Hertz in 1887 provide more details about radio communication.

---

[72] "Brain Waves: A theory", The Spectator, 01/30/1869.
http://ulsfmovie.org/docs_pd/Knowles_James_18690130.pdf

To come now to my crude hypothesis of a *Brain-Wave* as explanatory of them and of kindred stories.

Let it be granted that whensoever any action takes place in the brain, a chemical change of its substance takes place also; or, in other words, an atomic movement occurs; for all chemical change involves—perhaps consists in—a change in the relative positions of the constituent particles of the substance changed.

[An electric manifestation is the likeliest outcome of any such chemical change, whatever other manifestations may also occur.]

Let it be also granted that there is, diffused throughout all known space, and permeating the interspaces of all bodies, solid, fluid, or gaseous, an universal, impalpable, elastic "Ether," or material medium of surpassing and inconceivable tenuity.

[The undulations of this imponderable ether, if not of substances submerged in it, may probably prove to be light, magnetism, heat, &c.]

But if these two assumptions be granted, and the present condition of discovery seems to warrant them, should it not follow that no brain action can take place without creating a wave or undulation (whether electric or otherwise) in the ether; for the movement of any solid particle submerged in any such medium must create a wave?

If so, we should have as one result of brain action an undulation or wave in the circumambient, all-embracing ether,—we should have what I will call Brain-Waves proceeding from every brain when in action.

Each acting, thinking brain then would become a centre of undulations transmitted from it in all directions through space. Such undulations would vary in character and intensity in accordance with the varying nature and force of brain actions, *e.g.*, the thoughts of love or hate, of life or death, of murder or rescue, of consent or refusal, would each have its corresponding tone or intensity of brain action, and consequently of brain-wave (just as each passion has its corresponding tone of voice).

Why might not such undulations, when meeting with and falling upon duly sensitive substances, as if upon the sensitized paper of the photographer, produce impressions, dim portraits of thoughts, as undulations of light produce portraits of objects?

Figure 4.18. Partial text from "Brain Waves: A Theory" by James Knowles- makes sense to me too!

The only reason we know that this is James Knowles, for whom no portrait can be found, is from this article 30 years later: 'Wireless Telegraphy and "Brain Waves"'[73]. In this article the mysterious

---

[73] James Knowles, "The Twentieth Century", Volume 45, 1899, p858.
http://books.google.com/books?id=VAADAAAAIAAJ&pg=PA858

codeword "Potter" appears again adding to the feeling that "Potter" means something - maybe the name of a person who discovered some part of neuron reading or writing. These two articles are highly recommended reading for the excluded outsider - there are numerous descriptions and hints about what the insiders see in their eyes - in particular the neuron writing "suggesting" abuses done to outsiders. Is it not somewhat mysterious that this fact that all brains emit light with infrared and radio frequencies is not a basic part of our formal education and public scientific research even 140 years later? Should we not see low cost cameras and many movies that capture the many different invisible frequencies of light emitted from bodies by now?

1875

The Electric Currents of the Brain. By RICHARD CATON, M.D., Liverpool.—After a brief *résumé* of previous investigations, the author gave an account of his own experiments on the brains of the rabbit and the monkey. The following is a brief summary of the principal results. In every brain hitherto examined, the galvanometer has indicated the existence of electric currents. The external surface of the grey matter is usually positive in relation to the surface of a section through it. Feeble currents of varying direction pass through the multiplier when the electrodes are placed on two points of the external surface, or one electrode on the grey matter, and one on the surface of the skull. The electric currents of the grey matter appear to have a relation to its function. When any part of the grey matter is in a state of functional activity, its electric current usually exhibits negative variation. For example, on the areas shown by Dr. Ferrier to be related to rotation of the head and to mastication, negative variation of the current was observed to occur whenever those two acts respectively were performed. Impressions through the senses were found to influence the currents of certain areas; *e. g.*, the currents of that part of the rabbit's brain which Dr. Ferrier has shown to be related to movements of the eyelids, were found to be markedly influenced by stimulation of the opposite retina by light.

Figure 4.19. Article by Richard Caton, M.D. in 1875 describing the first publicly known direct neuron reading.

The earliest known "direct neuron reading" (the electricity in nerve cells measured) and the earliest published recording of sensory evoked electric potentials measured on the brain (fig. 4.19)[74] that I have found yet. By this time people had been applying electricity to neurons for centuries, but I have not yet found any earlier examples of reading the values of neurons. Shouldn't people publicly recognize and celebrate the first known reading of neurons? Note that this report reveals that eyelid motor neurons and light detecting neurons are located near each other.

1876

Figure 4.20. Alexander Graham Bell- did he record millions of thought phone calls? Ah who cares.

Alexander Graham Bell patents a telephone (fig. 4.20). Clearly the "Bell" company (now AT&T) and the other telecom companies would certainly have some role in the many flying dust-sized particle cameras and neuron reading and writing networks if such things exist. They own all of the wired network (the Internet), and so probably they would own and develop most of the wireless devices and network too. Wealthy people and governments would probably own and operate some part of the remote neuron reading and writing devices too. There must be tremendous value in hearing phone call

---

[74] Richard Caton, "The Electric Currents of the Brain", British Medical Journal, 1875, V2, p278.

recordings, and seeing digital movies of pretty women and important people inside their houses. Then, of course, seeing their thought images and hearing their thought sounds, and in particular making them do things the neuron owners desire by remotely writing and activating their neurons (literally remotely controlling them to the neuron owners wishes using a lot of, but not constant, remote neuron writing) must be of tremendous value.

The entire system is most likely wireless up to the phone company and government data storage buildings. The data storage probably makes the US Library of Congress look small, and may contain the thought images of people going back even to the 1300s – all being horded and kept from the majority of the public. It seems obvious that the governments of Earth must have secret "thought image and sound recordings" libraries that are inaccessible for the many millions of excluded people that funded and still fund them. We paid for all this, but don't get to see any of it. Maybe when the public is ready for total freedom of all information, the time will be ripe to start making thought-images and sounds public to all those who helped to fund capturing and storing them.

1881

FIG. 4.—Image as reproduced by Receiver.

Figure 4.21. Shelford Bidwell and the first publicly known image sent electronically.

Shelford Bidwell sends and prints photographic images electronically using selenium cells (fig. 4.21).[75] This is like a fax, photocopier, and even videophone - although apparently only working in black and white. This is the first conversion (encoding and decoding) of an image into an electronic signal. People were sending images in 1881, but only in the 2000s has video messaging *finally* become common for the public- although still no D2B yet!

It thus appeared that the experiments could not be performed in Berlin, and the apparatus was accordingly removed

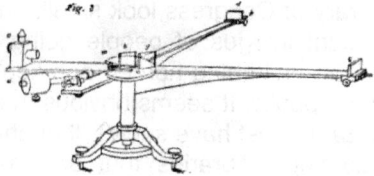

to the *Astrophysicalisches Observatorium* in Potsdam. Even here the ordinary stone piers did not suffice, and the apparatus

Figure 4.22. Albert Michelson and a drawing of the apparatus he used to measure and compare the speed of light in two directions at the same time.

Albert Michelson with Alexander Graham Bell's support initiates the end of the "light is a wave in an aether medium" theory by showing that light travels at the same speed with no change due to the movement of Earth in the supposed aetherial medium (fig. 4.22)[76]. Albert Einstein reinforces this end of aether in 1905.

---

[75] Shelford Bidwell, "Tele-Photography", Nature, Volume 23, Number 589, 10 February 1881, pp333-356.
http://www.nature.com/nature/journal/v23/n589/index.html
[76] Albert A. Michelson, "The relative motion of the Earth and the Luminiferous ether", The American Journal of Science, Volume 122, 1881, p120.
http://books.google.com/books?id=S_kQAAAAIAAJ

1885

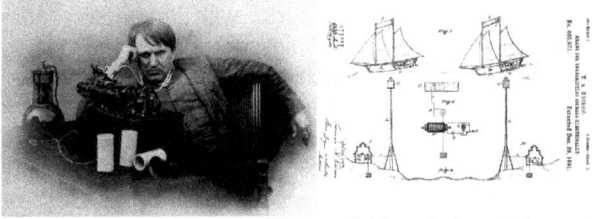

Figure 4.23. Thomas Edison and his wireless telegraph patent.

Invisible particle communication. Thomas Edison describes the first publicly known sending and receiving of text messages by invisible frequencies of light particles (wireless) (fig. 4.23).[77]

1887

Figure 4.24. Heinrich Hertz and a drawing from his paper.

Heinrich Hertz publicly explains "electrical resonance" (which allows specific ranges of frequencies of light particle beams to be filtered) (fig. 4.24).[78] This greatly popularizes the idea of low

---

[77] Edison patent 465,971, "Means for transmitting signals electrically".
http://www.google.com/patents/US465971?printsec=drawing#v=onepage&q&f=false
[78] H. Hertz, "Ueber sehr schnelle electrische Schwingungen", Annalen der Physik, Volume 267 (V. 31) Issue 7, March 1887, Pages 421 - 448.
http://books.google.com/books?id=WhY4AAAAMAAJ AND
http://de.wikisource.org/wiki/Benutzer:CK85/Untersuchungen_%C3%B

frequency particle (wireless) communication. Many people wrongly credit Hertz with inventing radio communication - Hertz's big contribution was going public with tuning in specific frequencies by electrical resonance and publicizing radio as a method of communication. It seems likely that Hertz, like Philip Reiss, the first to go public with the telephone, may have been murdered remotely with particles or microscopic devices for his telling the public about secret technology.

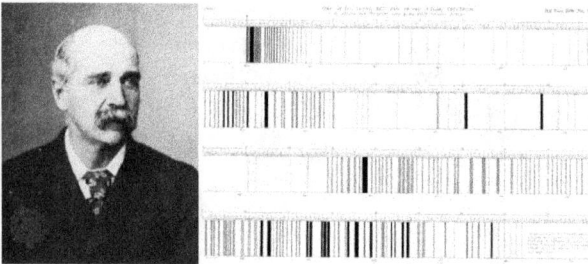

Figure 4.25. William Abney and a drawing from his paper.

William Abney (fig. 4.25) is the first to invent a photographic material (emulsion, a mixture of 2 chemicals) that works for infrared light (the silver compound turns black when contacted by light particles with infrared spacing).

Cber_die_Ausbreitung_der_elektrischen_Kraft_Kapitel_2
English Translation:
Heinrich Hertz, tr: D. E. Jones, "On Very Rapid Oscillations", Electric Waves, 1893, 1962, p29.
http://books.google.com/books?id=EJdAAAAAIAAJ

1889

Figure 4.26. The 1889 Bell symbol.

Alexander Graham Bell is in business connecting people's houses with wires and telephones. Does Bell start to record the phone call signals on their wires at this early date? Is this an inevitable characteristic of all phone companies around the Earth?

Figure 4.27. William Friese-Greene and the first (publicly known) films made on celluloid (1889-1890).

William Friese-Greene (fig. 4.27), invents the earliest public motion picture camera, describes capturing photographs from light emitted by the eye, and hypothesizes about capturing images from the eye from behind the eye.[79]

1895

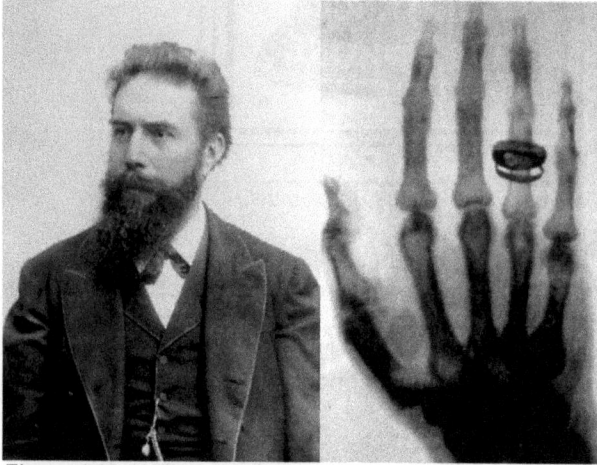

Figure 4.28. William Röntgen and an early X-ray photo.

Wilhelm Röntgen reports finding x-rays[80] (fig. 4.28).

---

[79] William Friese-Greene, "Photographs Made with the Eye", Photographic Times, 1889, p108-109.
http://books.google.com/books?id=-bUaAAAAYAAJ&pg=PA108
(and see the movie "The Magic Box")

[80] Wilhelm Conrad Röntgen, "Über eine neue Art von Strahlen", Aus den Sitzungsberichten der Würzburger Physik.-medic. Gesellschaft 1895.
http://de.wikisource.org/wiki/%C3%9Cber_eine_neue_Art_von_Strahl
English translation:
"On a New Kind of Rays", Nature, Volume 53, Number 1369, Jan. 23, 1896, p274.
http://books.google.com/books?id=nWojdmTmch0C&pg=PA274

1897

Figure 4.29. William Crookes, chemistry scientist, and hero of remote neuron reading and writing.

William Crookes (fig. 4.29) states publicly that x-rays could be possibly used in the science of telepathy, in addition to dropping numerous hints.

Crookes writes: "...It seems to me that in these rays {xrays} we may have a possible mode of transmitting intelligence, ...Let it be assumed that these rays, or rays even of higher frequency, can pass into the brain and act on some nervous centre there. ... In this way some, at least, of the phenomena of telepathy, and the transmission of intelligence from one ... to another through long distances, seem to come into the domain of law..."[81].

---

[81] William Crookes, "Address by the President", Proceedings of the Society for Psychical Research, V12, 1897, p338.
http://books.google.com/books?id=hBErAAAAYAAJ

Figure 4.30. The first public electric display and Ferdinand Braun.

The first electric display (cathode ray tube), by Ferdinand Braun[82] (fig 4.30).

1899

James Knowles reprints his "Brain-Waves: A Theory" article with a preface, and reveals his name as the original author of the 1869 article.

1900

The book "The Wonderful Wizard of Oz"[83] may be a metaphor for the confrontation that may occur if the public finally sees and uncovers all of the lies, tricks, and deceptions of those who own and control RNRAW, but it also serves as a metaphor for the public becoming wise to the lies of religions and religious authority, like Popes, Priests, Televangelists, etc. The book is made into a major movie in 1939 that contains the iconic phrase "pay no attention to the man behind the curtain!", which probably by no coincidence includes the letters "att" in the word "attention". Another interesting part in the book, is that everybody must wear green spectacles, which may hint at a "pre-public-D2B" system where excluded people can wear glasses (like the "Google Glass") with tiny cameras that move together with

[82] Ferdinand Braun, "Ueber ein Verfahren zur Demonstration und zum Studium des zeitlichen Verlaufes variabler Ströme", Annalen der Physik und Chemie, vol. lx., 1897, p. 552-559.
http://books.google.com/books?id=rXgMAAAAYAAJ&pg=PA464
[83] L. Frank Baum, "The Wonderful Wizard of Oz", 1900.
http://books.google.com/books?id=qbV65PabTEYC

the person's eye movement. In this way D2BW-like windows (for example money offers, and information about people and places around them) can be drawn to appear virtually in their surroundings, just like D2BW, but on the semitransparent LCD screens (in place of the "lens") in front of their eyes.

1901

Figure 4.31. William Herbert Rollins killed guinea pigs using x-rays, but humans would never do that to other humans.

William Herbert Rollins (fig. 4.31) kills guinea pigs with x-ray frequency light particles.[84]

1903
Edison makes the first motion picture available for the public: "The Great Train Robbery".

---

[84] William Rollins, "X-Light Kills", Boston Medical and Surgical Journal, February 14, 1901, p173.
http://books.google.com/books?id=0sUEAAAAYAAJ&pg=PA173

<u>1904</u>

Figure 4.32. "Hypnosis", by Sascha Schneider (1870-1927)

Sascha Schneider was a homosexual German painter who made the above painting. Homosexuality (of both men and women) is very likely a common excuse to exclude people, and many D2B suggestions on excluded people are to do homosexual activities. In "Hypnosis" (fig. 4.32) notice that, similar to looking at a bright light with the eyes closed, light can be drawn to the eye and seen even when the eye lid is shut. In fact, much of D2B may operate just by writing directly to the retina through the eye from the front with invisible frequencies of light (and to the ear and skin in a similar way). Note too that the hooded figure may be a theme to depict the neuron writers throughout the long history of those in the secret club of neuron reading and writing.

1909

Figure 4.33. Jean Perrin.

Earliest evidence of neuron reader and writer devices that are microscopic in size: Jean Perrin (fig. 4.33) talks about "dust" particles and uses the word "thought" three times in one paragraph - statistically unlikely to be coincidence. This is pretty solid evidence of dust-sized technology (wireless cameras, neuron readers and writers) already in 1909- *pre World War 1 and 2*. This is in a famous paper[85] about the Brownian motion. Perrin writes, talking about Brownian motion (translated into English):

"...The singular phenomenon discovered by Brown did not attract much attention. It remained, moreover, for a long time ignored by the majority of physicists, and it may be supposed that those who had heard of it **thought** it analogous to the movement of the **dust** particles, which can be seen

---

[85] Perrin, "Mouvement Brownien et Realite Moleculaire", Annales de chimie et de physique, S8, 09/1909, p3.
http://gallica.bnf.fr/ark:/12148/bpt6k349481.image.r=annales+de+chimie+et+de+physique.f3.langEN
Perrin, tr: Soddy, "Brownian Movement and Molecular Reality", London: Taylor and Francis (1910).
http://www.archive.org/details/brownianmovement00perr

dancing in a ray of sunlight, under the influence of feeble currents of air which set up small differences of pressure or temperature. When we reflect that this apparent explanation was able to satisfy even **thoughtful** minds, we ought the more to admire the acuteness of those physicists, who have recognised in this phenomenon, which they **thought** insignificant, a fundamental property of matter. "

May

Figure 4.34. Secret RNRAW hero Hugo Gernsbach and his 1909 "Electrical Experimenter" magazine cover with "The Thought Recorder".

The magazine "Electrical Experimenter", published by Hugo Gernsback, displays the above image (fig. 4.34) in May 1909. Ten years later the Syracuse Herald will run an expanded copy of this image (fig. 4.36).

1911
Hugo Gernsback publishes the earliest public description of a machine that can record thought-sounds in his story "Ralph 124C 41 + "[86] (fig. 4.35).

---

[86] Hugo Gernsback, "Ralph 124C 41 +", "Modern Electrics", Modern Electrics Publication, New York, Vol. 4, No. 3, June 1911. Taken from "Modern Electrics", Volume 3-4, Jan-Dec 1911, p165-168.

After a few minutes' reflection he pressed the button, and immediately a wave line, traced in ink, appeared on a narrow white fabric band, the latter resembling a telegraph recorder tape.

The band which moved rapidly, was unrolled from one reel and rolled up on another. Everytime 124C 41 wished to "write" down his thoughts, he would press the button, which started the mechanism as well as the recording tracer.

Below is shown the record of a Menograph, the piece of tape being actual size.

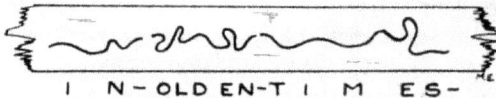

Where the waveline breaks, a new word or sentence commences; the three words shown are the result of the thought which expresses itself in the words, *"In olden times."*   .   .   .

The Menograph was one of 124C 41's earliest inventions, and entirely superseded the pen and pencil. Anyone can use the apparatus; all that is necessary to be done is to press the button when an idea is to be recorded and to release the button when one reflects and does not wish the thought-words recorded.

Figure 4.35. From the 1911 Hugo Gernsbach story "Ralph 124C 41 +". It seems likely that people were recording thought-audio in 1911- long before most of us were even born. ; - (

1919

Figure 4.36. 1919 Picture from the Syracuse "Herald" newspaper.

The Syracuse Herald features an adapted version of the 1909 image from Hugo Gernsbach's "Electrical Experimenter" magazine in an article titled "This Machine Records All Your Thoughts" (fig. 4.36).[87]

1922

George Emory Hale uses the word "render" in his book "The New Heavens"[88] which is one of many historical keywords to indicate that in 1922, the neuron reading and writing owners are modeling humans in 3D and in real-time. Probably by this time, the shape and location of most humans in developed nations are stored and tracked by the phone companies and governments of all major nations, but the public remains unaware. Even with this knowledge those in the know apparently do little to stop murder or even to expose or capture those who they saw commit murder.

---

[87] "This Machine Records All Your Thoughts", Syracuse Herald, 06/08/1919.
http://ulsfmovie.org/docs_pd/This_Machine_Records_All_Your_Thoughts_Syracuse_Herald_19190608.pdf
[88] George Ellery Hale, "The New Heavens", 1922, p27.
http://books.google.com/books?id=bx0SAAAAYAAJ

1923

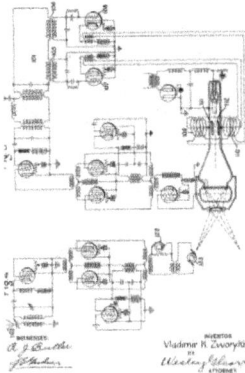

Figure 4.37. An "electronic image" camera and Vladimir Zworykin.

Vladimir Zworykin goes public with an electronic image camera and display (a television camera and cathode-ray tube display)[89] (fig. 4.37).

1927

Figure 4.38. Image from AT&T's 1927 movie "That Little Big Fellow".

---

[89] Vladimir K. Zworykin's patent 2141059, Filed: Dec 29, 1923.
http://www.google.com/patents/about?id=bdYBAAAAEBAJ

AT&T releases the movie "That Little Big Fellow" (fig. 4.38), a movie that contains a picture of the thought-screen of a messenger with an image of "the thinker" statue on it. This is clear evidence that neuron reading and writing was developed by 1927 – note: *before World War 2*. The thinker uses his right arm, while the delivery guy uses his left arm – perhaps reflecting a "right and left" analogy.

Note also that the bizarre US copyright law forbids people from publishing most text, sounds, and images created past 1923 except as the nebulous "fair use", while the neuron owners sit back and *get paid* money by D2B consumers to see all of our latest in-home activities, and our thought-images and thought-sounds directly in front of their eyes without getting permission to copy and distribute any of our works, or paying, we, the subjects of all the videos, a cent.

1928

Figure 4.39. Disney's "Steamboat Willie" may hint at how a person looks with "eye" and "thought" screen circular windows.

"Mickey Mouse" is shown publicly. Believe it or not, this is evidence for remote neuron reading and direct-to-brain windows as early as 1928. The ears of Mickey and Minnie Mouse (fig. 4.39) might represent how an "eye" and "thought" window look. Even the three circles (the head and ears) may represent what a D2B consumer sees in their eyes for each person they are looking at: a circle with the front-view of the person from a dust-sized microcamera floating nearby, plus two more circles for their thought and eye screen attached on top, because the thought and eye screen might not be large enough in the first circle- but perhaps there is just one circle for all three. They see a front view of the subject's torso, a circular window of what the subject's eyes see, and another circular window showing any image the subject is thinking of.

Seeing these two circles in their most common position (above the head) on Mickey and Minnie Mouse, can be viewed as if we are looking at a black haired person with a very close (although blank) thought and eye screen above them. This image may have provided relief to some neuron consumers. They must have thought, after WW1 and all the neuron abuse and lies they saw - "now here is some hope - seeing eyes and ears will probably go public within 10 years...", but how wrong and inaccurate that false hope has proven to be. Ownership of neuron writing, perhaps by its nature, has caused a shocking stagnation that persists - like religious myths - for possibly thousands of years.

Note that "MM" for "Mickey Mouse" has a lot of significance as an abbreviation for "mass murder", "muscle mover", "muscle molestation", and the upside-down WW of "World War".

Looking at this timeline, you can see that the Disney company has hinted about RNRAW more than most other companies.

Figure 4.40. "Popular Science" publishes an article "Is Telepathy All Bunk?". Some wealthy people were clearly bucking the secret system in the early 1900s.

Popular Science prints an article titled "Is Telepathy All Bunk?"[90] (fig. 4.40) Thomas Edison famously said "Religion is all bunk"- possibly an association to, or a story funded by Edison.

1929

Abb. 4. 40 jähriger Mann. Große linksseitige, von der Stirn bis in die Parietalgegend reichende Knochenlücke. Doppelspulengalvanometer. Kondensation. Nadelelektroden anbentan im Bereich der Knochenlücke, 4.5 cm voneinander entfernt. Oben Schwankungen der epidural abgeleiteten Kurve, unten Zeit in ¹/₁₀ Sekunden.

Figure 4.41. Hans Berger and variations in electric voltage as measured on the skull, from his 1929 paper.

German psychiatrist, Hans Berger, applies electrodes to the human skull and connects the other end to an oscillograph, which records the changes in electric potential (voltage) (fig. 4.41). From these signals Berger distinguishes and names "alpha" and "beta" waves. This work begins the

---

[90] Kenneth Wilcox Payne, "Is Telepathy All Bunk? What Scientists Have Discovered About This Widely Discussed Subject in Thousands of Exhaustive Tests", 02/1928, p32.
http://books.google.com/books?id=VycDAAAAMBAJ&pg=PA32

science of electroencephalography, which is useful in diagnosing epilepsy.

Was Berger excluded from neuron reading and writing, or a D2B consumer that advanced public knowledge? Even today, this comparatively primitive encephalograph technology is viewed as state of the art, and is being sold for use in video games as a new and modern device - where humans control objects by relaxing and tensing their mind, or by using different parts of their mind, very far from the modern neuron reading and writing done with nanotechnology. Berger was probably murdered by the Nazis in 1941.

1930

First public description of a handheld ray weapon "The Black Star Passes", by John W. Campbell.[91]

Figure 4.42. Upton Sinclair and William McDougall, are big RNRAW hinters- but don't talk explicitly about RNRAW done with technology.

---

[91] John Campbell, "The Black Star Passes", Experimenter Publications, 1930.
http://www.gutenberg.org/files/20707/20707-h/20707-h.htm

The famous "muckracker" Upton Sinclair (fig. 4.42) publishes "Mental Radio"[92] about telepathy. The German version has a preface written by Albert Einstein. William McDougall (fig. 4.42) of Duke University also writes a preface. McDougall writes (words that might be directed at those who continue the RNRAW lie): "...are grossly stupid, incompetent and careless persons or have deliberately entered upon a conspiracy to deceive the public in a most heartless and reprehensible fashion." and "...science furnishes us no good reasons for denying that its activity {the mind} may affect another mind in some fashion utterly obscure to us ... For we do seem to know with very fair completeness the possibilities of influence extending from the printed word to the experimenter; and under the conditions all such possibilities seem surely **excluded**. ...".

Sinclair writes: "TELEPATHY, or mind-reading: that is to say, can one human mind communicate with another human mind, except by the sense channels ordinarily known and used-seeing, hearing, feeling, tasting and touching? Can a thought or image in one mind be sent directly to another mind and there reproduced and recognized? If this can be done, how is it done? Is it some kind of vibration, going out from the brain, like radio broadcasting?". Note that the word "reproduced" probably hints at the excludocidal nature of the D2BW lie- how denial of basic D2BW service is used to "weed out" many honest, non-religious, non-white, poor, etc. family lines of humans. Note too that for years people have simply used the word "thought" without distinguishing "thought-audio" and "thought-images", and that has probably added to the confusion and doubts of the poor excluded people in their understanding of what thought actually is.

---

[92] Upton Sinclair, "Mental Radio", 1930.
http://books.google.com/books?id=4sbmCMiXmo8C

1933

Figure 4.43. Image of "thought-projector" projecting Tesla's thought-images.

"The Deseret News" publishes an article with the above image (fig. 4.43) in which Nikola Tesla is quoted as saying: "I expect to photograph thoughts"[93] . In a 07/11/1934 New York Times article Tesla talks about a death ray that can destroy planes and millions of people in an instant without leaving any trace[94], and in another article on 07/11/1935 rejects the theory of relativity as being "a mass of error and deceptive ideas violently opposed to the teachings of great men of science of the past and even to common sense."[95].

[93] Carol Bird, "Tremendous NEW POWER soon to be released", The Deseret News, Salt Lake City, Utah, 9/9/1933.
http://news.google.com/newspapers?nid=336&dat=19330909&id=0KVOAAAAIBAJ&sjid=9bUDAAAAIBAJ&pg=6908,2324471
[94] "Tesla, at 78, Bares New 'Death-Beam'", New York Times, July 11, 1934, p. 18, c. 1
http://www.tesla-coil-builder.com/Articles/jul_11_1934a.htm
[95] "Tesla, 79, Promises to Transmit Force", New York Times, July 11, 1935, p 23, c.8
http://www.tesla-coil-builder.com/Articles/jul_11_1935b.htm

1936

Figure 4.44. Joseph Banks Rhine, hints about RNRAW indirectly, like Sinclair, but unlike Crookes, without the assistance of electronic particle devices.

Joseph Banks Rhine (fig. 4.44) coins the phase "Extra-sensory perception" (ESP) in his book "Extra-sensory Perception"[96]. Like Upton Sinclair's "Mental Radio" of 1930, William McDougall (fig. 4.41) of Duke University writes a foreword. McDougall writes: "...the author...will pardon my **intrusion on his privacy**. ...Indeed in this age when we erect monuments to the boll-weevil, send up prayers for drought, pest and plague, and are chiefly concerned to make one ear of wheat grow where two grew before, it is difficult to retain enthusiasm for botanical research, unless one is a scientist of the peculiarly **inhuman** type.".

---

[96] J. Rhine, "Extra-sensory Perception", Bruce Humphries, Boston, 1935.

Figure 4.45. Image from the short "Popeye" movie "Hold the Wire". This "identity theft" is a common abuse by the telecoms.

Paramount releases the short Popeye movie "Hold the Wire"[97] (fig. 4.45) in which Bluto intercepts the phone line and pretends to be Popeye in order to ruin the relationship between Popeye and Olive Oil. This "identity theft", to cause conflict within the "enemy" (usually the educated and honest), is typical of the neuron writing deception and the bizarre, un-democratic, un-observable, and un-punishable communication service of Earth.

---

[97] "Hold the Wire", Paramount Pictures, 1936.
http://www.youtube.com/watch?v=LoVW7B2JDaM

1937

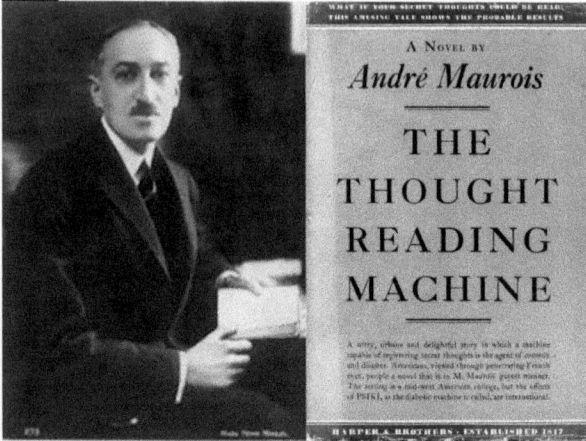

Figure 4.46. André Maurois and his book "The Thought Reading Machine" (1937).

André Maurois writes "La Machine à lire les pensées" in French, which is published in English as "The Thought-Reading Machine"[98] (fig. 4.46). The book describes a scientist at a university that invents a device that can record the sounds people think, from their larynx. It is an unbelievably good book for helping to warn and inform excluded people about the RNRAW secret. Although it is fiction, it may relate to the real story (kept secret for perhaps 700 years) of the inventions of seeing and hearing thought. The book also contains transcripts of what thought audio sounds like; for example how the wife of the main character becomes distracted while reading a book, and her thought-audio changes from reading the text of the book to other thought-audio about her personal relationships.

1940

Franklin Delano Roosevelt, President of the United States says, "This nation will remain a neutral

---

[98] André Maurois, "La Machine à lire les pensées", Gallimard, 1937
English translation: "The Thought Reading machine", Harper, 1938.

nation, but I cannot ask that every American remain neutral in **thought** as well. Even a neutral has a right to take account of facts. Even a neutral cannot be asked to close his **mind** or close his conscience.".[99]

1944

Figure 4.47. Image from "Plane Daffy" of a videophone- at the time, not yet available to the public.

Warner Brothers releases the cartoon "Plane Daffy"[100] (fig. 4.47) which shows Hitler, Göring, and Goebbels using a video phone - but this is decades before something as simple as a video phone is made available to the public.

[99] FDR, Fireside Chat (on the Outbreak of World War II), Washington D.C., 09/03/1939.
http://books.google.com/books?id=Whu3zcbjqvkC&pg=PA68
[100] "Plane Daffy", Warner Brothers, 1944.
http://youtube.com/watch?v=SEb0Z6cbe8c

1947

Figure 4.48. Image from Disney's "Delayed Date" reminding us that many other species also have an eye and thought screen, and thought audio too.

Disney's Mickey Mouse short film "Mickey's Delayed Date"[101] (fig. 4.48) shows a thought-screen over the dog Pluto. Just another shocking truth denied to excluded, the pleasure of seeing and hearing the thoughts of all the other species- nice D2B owners and consumers huh?

1954

THE MEMBRANE CHANGE PRODUCED BY THE NEUROMUSCULAR TRANSMITTER

By J. del CASTILLO and B. KATZ

From the Department of Biophysics, University College London

(*Received 30 March 1954*)

Until recently, it was generally believed that the action potential which a nerve impulse sets up in a muscle fibre is identical with that produced by direct stimulation. Recent work has shown that this is only true if the impulse is recorded at a point remote from the neuromuscular junction. At the end-plate itself, the amplitude of the spike initiated by the synaptic transmitter is smaller than that due to a direct electric stimulus (Fatt & Katz, 1951, 1952 *a*, *b*;

Figure 4.49. After decades of neuron research all we get is: "remote", notice the comical citation of "Fatt & Katz".

---

[101] "Delayed Date", Disney, 1947.
http://www.youtube.com/watch?v=6UaYGygiJNE

German-British physiologist and Nobel Prize winner Bernhard Katz uses the word "remote" in the first paragraph of a paper about electrically making neurons fire ("direct neuron writing")[102] (fig. 4.49). Gee, do you think remote muscle contraction might be a productive line of research....naw!

Katz had used the word "indirect" before this as a double meaning- the first meaning being electrical stimulation of the nerve which makes the muscle contract, as opposed to making a muscle contract by applying an electric current to the actual muscle directly.

1955

"The Seven Year Itch" - the famous movie where Marilyn Monroe's dress is blown up by air from a subway tunnel. The title could be a play on "700 years of the remotely caused itch" - if for the year 2000 - that would put neuron writing in the 1300s but for the 1900s it would be the 1200s. It's a stretch, but very well may be- 700 or more years allows for a lot of time to devote to creative hinting.

1958

Cordwainer Smith publishes "No, No, Not Rogov!"[103] in which Stalin has a thought reading and writing machine built that is used to send thoughts to make other people confused, to commit suicide, etc. Many times people say "bear in mind" which may mean that "remember that Russia has had remote neuron reading and writing for many centuries". The only people that are not in on the D2BW secret, the "big enemy", are most of the poor, lawful, honest and decent public, who are routinely tricked and misled with "false-flag" operations and false stories.

---

[102] del Castillo J, Katz B, "The membrane change produced by the neuromuscular transmitter", The Journal of Physiology, Vol. 125, No. 3. (28 September 1954), pp. 546-565.
http://jp.physoc.org/content/125/3/546.full.pdf
[103] Cordwainer Smith, "No, No, Not Rogov!", Quinn Publishing, 1958.
http://www.baenebooks.com/chapters/1416521461/1416521461___1.ht
m

Max Knoll and team find that light patterns can be experienced when a small voltage is applied by two electrodes on different parts of the human face, and the voltage oscillated in the encephalographic frequency range.[104]

1959

Device inside body controlled remotely: radio controlled artificial pacemaker.[105] This is the first publicly known extension of Galvani's 1791 remote ("wireless" or "particle") neuron writing. But note that this device is not "intracellular" (inside a cell). Pacemakers continue to get smaller, and the pacemaker may perhaps eventually be the first remotely controlled "neuron organelle", inside individual heart muscle or nerve cells, but perhaps intracellular devices to improve hearing or vision will be the first.

1960

John F. Kennedy states: "Can a nation organized and governed such as ours endure? That is the real question. Have we the nerve and the will? Can we carry through in an age where we will witness not only new breakthroughs in weapons of destruction-- but also a race for mastery of the sky and the rain, the ocean and the tides, the far side of space and **the inside of men's minds**?".[106]

---

[104] M. Knoll, J. Kugler, "Subjective Light Pattern Spectroscopy in the Encephalographic Frequency Range", Nature, V184, N4701, 12/05/1959, p1823-1824.
http://www.nature.com/nature/journal/v184/n4701/pdf/1841823a0.pdf
[105] Glen, Mauro, Longo, Lavietes, Mackay, "Remote stimulation of the heart by radiofrequency transmission", New England Journal of Medicine, v261, 11/5/1959, p948-51.
http://www.nejm.org/doi/pdf/10.1056/NEJM195911052611905
[106] John F. Kennedy Accepts the Democratic Nomination for President, Los Angeles, July 14, 1960.

1964

Figure 4.50. A "Thought Projector Helmet" shown in a "Fantastic Four" comic book.

A "Fantastic Four" comic book[107] shows a "Thought Projector Helmet" (fig. 4.50) which shows thought-images. Notice that, aside from images of food, nude and sex-related thought-images are probably commonly seen on thought-screens- in particular when a person (or even other species) is masturbating or having sex.

---

[107] "Fantastic Four" Issue 27, June, 1964.

1967

Figure 4.51. Image from "The President's Analyst" (1967).

"The President's Analyst"[108] (fig. 4.51) hints at how the phone company may be involved in microscopic technology, widespread surveillance, and thought seeing, hearing, and sending devices. This movie also contains, perhaps the most explicit and accurate public description of remote neuron reading and writing to date, including even an inside-the-body device (microchip) that makes calling people through thought possible (presumably using thought-audio and/or thought-images- a confusing point probably for many excluded people who don't realize that thought is made mostly of sounds and images, and that their thought-audio can be heard and thought-images seen externally by many people because of this still-secret technology) – and then from a phone company representative no less.

1968

January 31

Around 17 minutes into an interview on the "Tonight Show with Johnny Carson", Jim Garrison (hero of the *still* unofficial truth about the JFK murder) says "...I can't look into their brains, Johnny, and tell you why they did it", and Johnny Carson

---

[108]Theodore J. Flicker , "The President's Analyst", Paramount Pictures, 1967

replies with "...Now you expect somebody to be galvanized into action..."; Garrison hinting about the massive secret of seeing and hearing eye, thought-screen, ear, and thought-sound recordings, and Carson hinting about remote muscle moving - i.e. what Galvani did (and perhaps remote muscle murder).

1975

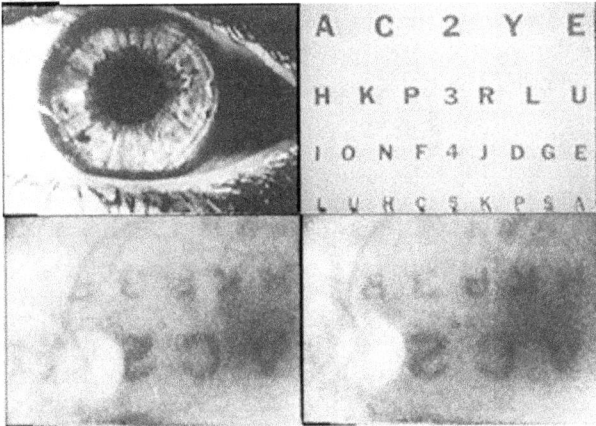

Figure 4.52. Image from "The Incredible Machine" (1975)

The   PBS television program "The Incredible Machine" (fig. 4.52) by National Geographic shows a sequence of images that contains nearly 5 seconds of an eye chart on the back of the retina of an eyeball using an opthalmoscope, perhaps the first time the public has seen this. In addition, a person moves a train using "alpha" brain waves.

early 1980s
A Head and SHoulders television ad slogan is "Because that little itch should be telling you something" which hints at the most prolific career molesters to ever live free on Earth, who cause lawful people to itch using microscopic light particle devices hovering in almost every square centimeter of Earth. (Notice how remote neuron writing made the above "SH" capital on "Head and Shoulders"- to

spell "SH" (i.e. "be quiet"). Usually I correct these constant molestations but I am leaving it this time.) Kudos to Head and SHoulders. Yes, that itch should be mf'in telling us something: that something is very, very wrong and very rotten with our phony phone-company laser undermocratic government. (funny, somebody made me type dermocratic instead of democratic- now that is a fast thinking computer). When the remote neuron writers do a typo – they go the full typo – remotely moving the actual muscle to make the typo- not just changing the letter on the computer electronically.

1985

In the PBS television series "Cosmos", Sagan says "we, ourselves, are far from decoding what a brain thinks, but in the future, it may well be possible....a disquieting prospect". And "Across the centuries an author is speaking to you, clearly in your head."

Fig. 5. SEM photograph of a 24-step stepping micromotor. The gap between the stator and rotor is $2 \mu m$.

Figure 4.53. Image of microscopic motor

Microscopic motor. Fan and team at the University of California at Berkeley publish the first public image of a microscopic electromagnetic motor (Fig 4.53).[109]

1990

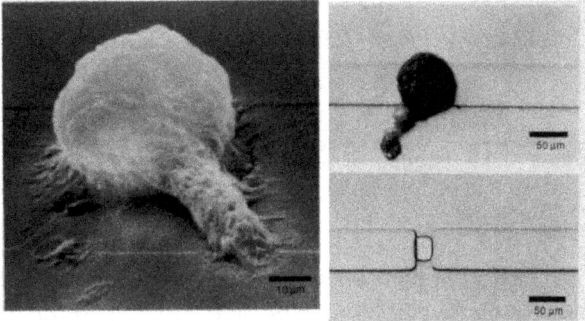

Figure 4.54. Images of a neuron from the Fromherz, et al. paper.

Fromherz and team interface a neuron with a field-effect transistor (fig. 4.54). They measure signals that pass through the neuron.[110]

---

[109] Long-Sheng Fan; Yu-Chong Tai; R.S. Muller; , "IC-processed electrostatic micro-motors," Electron Devices Meeting, 1988. IEDM '88. Technical Digest., International , vol., no., pp.666-669, 1988.
http://ieeexplore.ieee.org/stamp/stamp.jsp?tp=&arnumber=32901&isnumber=1415
[110] P Fromherz, A Offenhausser, T Vetter, and J Weis, "A neuron-silicon junction: a Retzius cell of the leech on an insulated-gate field-effect transistor", Science 31 May 1991: 252 (5010), 1290-1293.
http://www.sciencemag.org/content/252/5010/1290.short

1999

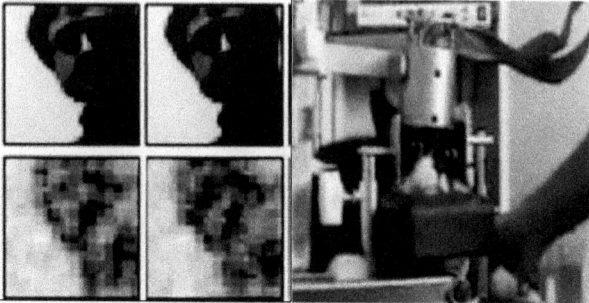

Figure 4.55. Images of what a cat's eye sees, and the apparatus used to capture the images.

UC Berkeley employee Yang Dan and team captured these images by connecting 177 electrodes to the thalamus of a cat (fig. 4.55).[111] Much of this research would be unnecessary if the D2BW owners, presumably AT&T, would simply go public with D2BW.

2000

In the movie "What Women Want" Mel Gibson is accidentally electrocuted and can hear thoughts of people around him. Gibson then uses this ability to deliver people exactly what they want. This is really an amazing phenomenon that is actually happening in reality because people are keeping neuron reading a secret and have for 200+ years apparently. A person that can hear thought can easily "lay out" by knowing exactly what the outsider (or even fellow insider) is expecting and wanting, while those victims who are excluded from hearing thought cannot possibly compete and so don't reproduce; they are being selected out of existence by the neuron reading and writing "selection" that

---

[111] Garrett B. Stanley, Fei F. Li, and Yang Dan, "Reconstruction of Natural Scenes from Ensemble Responses in the Lateral Geniculate Nucleus", The Journal of Neuroscience, September 15, 1999, 19(18):8036-8042.
http://www.jneurosci.org/cgi/content/full/19/18/8036

continues to this day without the vast majority of humans even knowing it. This does not even account for the God-like ability to write suggestions in the form of fast images and sounds to the neurons of those D2BW outsiders, to make them say and do all the wrong things to potential mates, while steering their potential mate towards somebody else. It raises the question: how many D2B outsiders are in a relationship with a D2B consumer who used thought seeing and hearing to attract them? How will the D2B outsiders react if ever they find out? One trailer for this movie uses the ominous excludocidal keyword "cuckoo".

2001

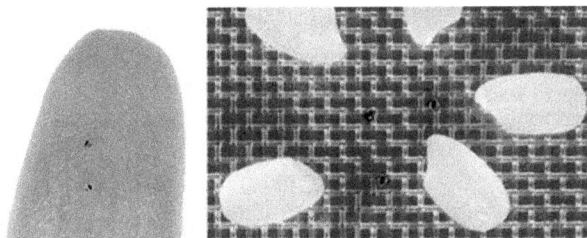

A New RFID with Embedded Antenna μ-Chip

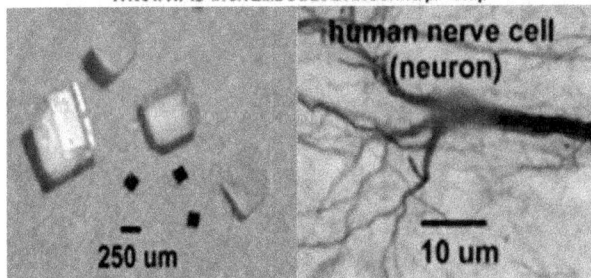

human nerve cell
(neuron)

250 um   10 um

Figure 4.56. Images of Hitachi's "μ-Chip"- most people don't know that wireless electronic devices can be this small (and smaller).

Hitachi sells the smallest and thinnest publicly known RFID (Radio frequency Identification) Integrated Circuit, the "μ-Chip" ("mYU"-chip) (fig.

4.56). These devices receive and transmit light particles.

In 2003 Hitachi reduces the size of these RFID chips to 50x50 micrometers. So these devices being made public is a key step in going public with the first "human made organelle"; a device that an external device can remotely signal to make an individual neuron, with its own specific MAC address, fire. To the naked eye the chips look like pieces of dust. Probably these chips could be easily inhaled or swallowed and enter a body that way.

2004

Cyril Wecht is quoted as saying "I will transmit my thoughts to {recent gun shot victim Taiwan President Chen Shui Bian}", hinting that thought audio and images can now be sent like any other audio or jpg/mpg images. I'm sorry to say that I cannot find the original source to cite. But it raises a good point that there are people who are more honest and those who are less honest, and I pride myself on being very honest in everything I say and do- I'm not looking to trick people- and there is no use lying when many people can see and hear your thoughts and your every activity. But how terrible the excluded are at recognizing the dishonest! The many liars about the JFK, RFK and 9/11 murders are clear everyday examples. We should trust those who consistently tell the truth and not those who consistently lie.

In addition, you can see that just because many people have been seeing, hearing and writing thought images and sounds for many centuries, doesn't mean that they are civilized and non-violent. In fact, almost the exact opposite is true; many apparently have become even more violent and barbaric. Then to make matters worse, they are more effective at violence because of their extreme technological advantage over those who do not control any RNRAW devices.

"The Final Cut", with Robin Williams, shows the perspective of "seeing eyes"; seeing what people see (the "eye screen", where images from the external universe are recorded in the brain). In the movie, a "Zoe chip" is implanted in people's brains. The image a D2B excluded person might get is of a big chip, but probably in reality there are many millions of chips and they are very small (neuron size). This movie represents a small step forward in explaining to the excluded public just a tiny part of the secret possible 700 year history of seeing and sending images, sounds, smells, and muscle moves remotely, and the shocking "exclusive society" of the D2B owners and consumers who get many preferred benefits, and who casually watch and hear everybody's thoughts without millions knowing that such a thing is even possible.

2007
July 25
Ted Huntington is on the KROQ "Kevin and Bean Show"[112] and they talk about seeing, hearing, and sending thought images and sounds.

2008

Figure 4.57. Image of people watching semi-transparent windows from Disney's "WALL·E".

The Disney-Pixar movie " WALL·E" shows people watching "menu" screens in front of their faces projected by their chairs (fig. 4.57) which looks very

---

[112] http://tedhuntington.com/kevinandbean.htm

similar to D2BW (and the famous R2D2 projection in the 1977 movie "Star Wars"). Notice that one woman is looking at 4 or 5 video squares of other people, perhaps a videoconference. This is a pretty solid hint about D2B windows, and again from the company that was founded on a character that shows the public the circular thought and eye screens above every head. This may be the first publicly known rendered drawing of an image that looks like semi-transparent D2B windows, but is not an actual 3D rendered image of D2B windows (like the image on the front cover of this book) which I may have been the first to make public around 2010- not to brag – but to express shock and disappointment. Notice that the chair beam is like mine but a few centimeters to the right. It seems impossible to project an image onto empty space- it only makes sense if the beam is on the brain making images only appear to be there.

December 10

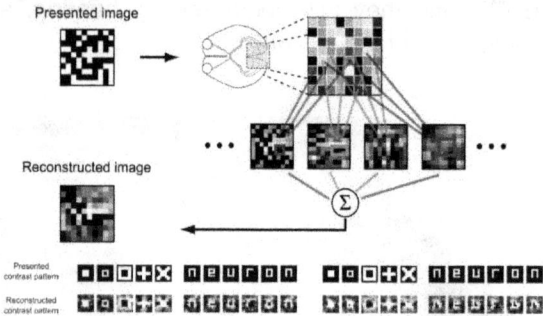

Figure 4.58. Image of remote neuron reading from Miyawaki, et al.

The first image captured by remote neuron reading is made public. Scientists in Japan, Miyawaki, et al. publish images that a brain sees using fMRI (functional magnetic resonance imaging) (fig. 4.58). This is the first public image of "eyes" (remotely captured images of what a brain sees) that I am aware of, although it very well may be that the first actual images captured of what a brain sees may date back as far as 700 years ago to the time of the first major universities in Europe. This is perhaps the best piece of physical evidence that proves that what a brain sees can be seen without having to touch the brain. One of the authors, Kamitani, told me in an email that they are able to distinguish between sounds of thought too, for example, they can distinguish between thought sounds like "po" and "go".

2010
I freely distribute 37 new original songs on the Internet, many about remote neuron reading and writing. Songs include: "Neuron Writing", "Where's Our Video?", "Let the People Get to See".[113]

November 12

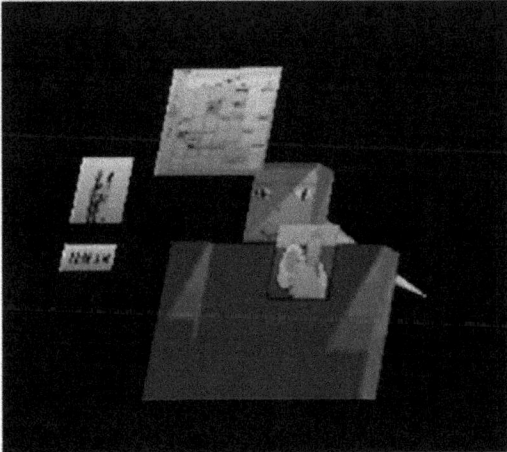

Figure 4.59. Possibly the first public image explicitly showing direct-to-brain windows.

---

[113] http://www.tedhuntington.com/songs.htm

I, Ted Huntington, make public the first (that I am aware of) public video showing direct-to-brain windows[114] (fig. 4.59). I learned how to render in 3D when I worked for a computer game company.

2011

Figure 4.60. "Scientific American Mind" cover shows a woman with possible D2B windows.

"Scientific American Mind"[115] publishes a photo of a woman with 3 windows of people that look similar to direct-to-brain "thought-screen" windows (fig. 4.60).

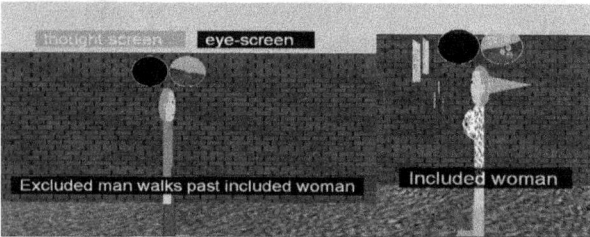

Figure 4.61. Image from free Internet video "Excluded man walks past included woman".

I make and freely distribute videos of what D2BW with a thought and (the first public) eye window may look and sound like in my short movie "Excluded man walks past included woman"[116] (fig. 4.61).

---

[114] http://www.youtube.com/watch?v=k1R5y8IMGk8
[115] Scientific Mind, March 2011, cover.
http://www.scientificamerican.com/sciammind/?contents=2011-03
[116] http://www.youtube.com/watch?v=r52Qdi-yJs4

<u>July 7</u>

Facebook announces a new free video call feature. Facebook engineer Philip Su states "If it was any easier than that one click, it would be reading your mind"[117]; which hints about the grotesque apartheid between those who receive direct-to-brain video windows and those who are excluded.

<u>August 16</u>

Figure 4.62. Image from a Carefree ad of a woman thinking about kissing a guy on her thought-screen.

A Carefree panty liner ad[118] (fig. 4.62) has a woman with a thought-screen. The growing quantity of these kinds of ads implies that many direct-to-brain windows consumers are repulsed by the segregation and neuron-writing abuse they must see and experience, and that possibly there is a growing group of terribly abused people who are becoming more aware that they are excluded from seeing thought-screens and hearing thought-audio. Companies are positioning themselves for a future where many excluded find out the truth, perhaps to make the case that they made an effort to inform the many D2B excluded victims, where those who stay

---

[117] Jessica Guynn, "Facebook unveils video chat with Skype", LA Times, 07/07/2011.
[118] http://www.tedhuntington.com/Carefree_Thought_Screen_2011.mp4

silent may be viewed as collaborators and accessories of the many unpunished D2B crimes.

September 22

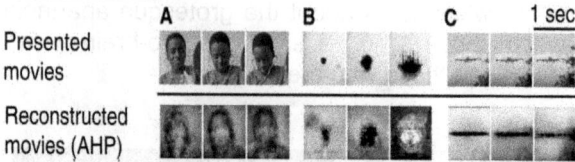

Figure 4.63. Remote neuron reading movies from Nishimoto et al.

Remote neuron reading of movies: One step beyond the 2008 Kamitani team still "neuron" image. Gallant and team at Berkeley have reconstructed movies seen by people remotely from behind their brain using fMRI (fig. 4.63).[119] As usual, we can mostly thank the education establishment for bringing information about remote neuron reading and writing to the public - not the corporate, government, major media, or religious establishment.

Advances are happening faster and faster which implies that remote neuron reading and writing and direct-to-brain windows can't possibly remain an unpublished secret by 2100 and probably more likely by 2050. Key steps are:

1) remote recording of ear and thought sounds
2) Images of thought captured remotely
3) RFID organelle (embedded in a neuron)– much better and precise remote neuron reading and writing.

Schiller and team at MIT recently published that direct-to-neuron electrical stimulation can produce images in monkeys[120]: In their abstract, which is

[119] Shinji Nishimoto, An T. Vu, Thomas Naselaris, Yuval Benjamini, Bin Yu, Jack L. Gallant, "Reconstructing Visual Experiences from Brain Activity Evoked by Natural Movies", Current Biology, Available online 22 September 2011, ISSN 0960-9822, 10.1016/j.cub.2011.08.031. http://www.sciencedirect.com/science/article/pii/S0960982211009377

written for average people and not immersed in techno-jargon, they compare it to a cochlear implant but for sight. Notice the similarity of the name "Schiller" with "shill" – when direct-to-brain consumers take money to say misleading and abusive statements out loud to excluded people; a disgusting part of the advanced segregation that has resulted from the hording for centuries of remote neuron reading and writing technology.

Galvani lit the spark publicly in the late 1700s and now all that is needed is to reduce the size of the electrodes and communicate with them using particles (wirelessly) through the skull, and that is the basic form of the secret modern remote neuron reading and writing which is used every 5 seconds by unseen criminals to contract some annoying muscle or fire some annoying neuron of ours.

By now there is a rapid growth and development of seeing eyes, seeing thought, hearing thought, sending images and sounds to and from brains, etc. Probably the dust-sized cameras, microphones, and neuron reading and writing particle communication devices are installed all over the earth and even perhaps out around other planets, moons and stars. And all the data they have collected over the centuries will serve as a wonderful resource once it is all made available for the public to see. After the initial shock, the public will be fascinated at seeing what their parents thought, and even what their great-great-great-grandparent's thought, seeing their daily lives, etc.

---

[120] Peter H. Schiller, et al, "New methods devised specify the size and color of the spots monkeys see when striate cortex (area V1) is electrically stimulated" PNAS 2011 ; October 10, 2011.
http://www.pnas.org/content/early/2011/10/04/1108337108.abstract?sid=93f53a7d-5e5b-4479-8a36-6071f2dd5fb0

We can even play this timeline forward into the future with expected technology going public or being invented:

2014

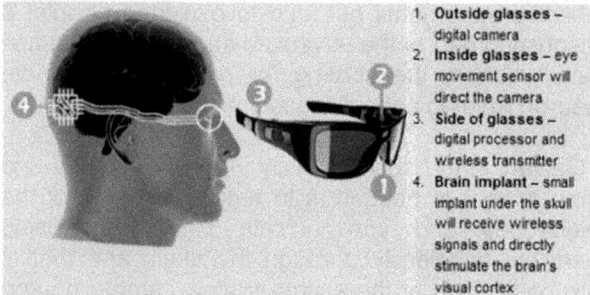

Figure 4.64. Monash University's "Direct to brain bionic eye"

Monash University plans to patent a "Direct to brain bionic eye": "a small implant under the skull will receive wireless signals and directly stimulate the brain's visual cortex"[121] (fig. 4.64). This form of extra-cellular electronic device assisted remote neuron writing will allow those without sight to see.

2015
Sound a brain hears recorded remotely.
Microscopic camera.

2018
Radio device functions as cell organelle.

---

[121] http://www.monash.edu.au/bioniceye/

2020

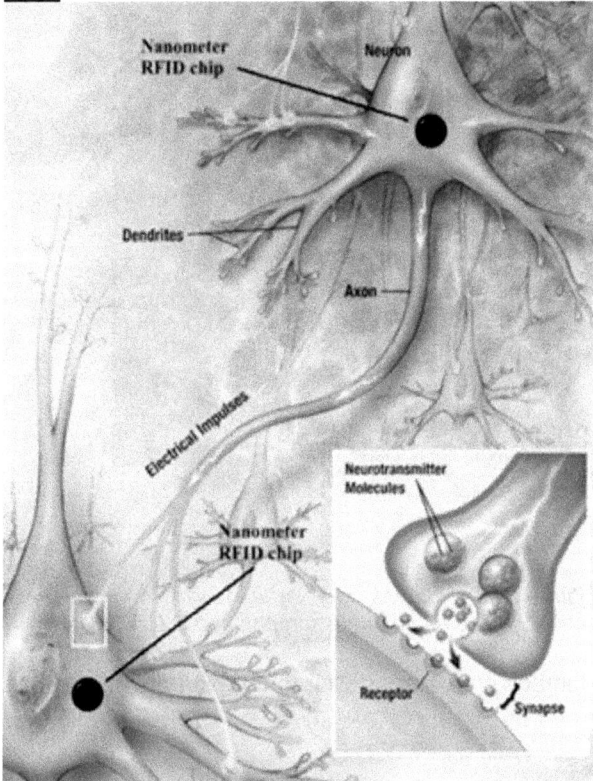

Figure 4.65. Perhaps what an RFID device may look like in a neuron.

Remote neuron writing using intracellular microscopic devices in neurons (fig. 4.65).

2025
Thought-images seen.
Thought-audio recorded and played out loud. Humans start to communicate publicly by thought image and sound only.
Microscopic flying camera.

2040
Artificial muscle walking robot.

2050

Figure 4.66. Humans walk around with robot servants, notice the eye and thought screens and D2B windows.

Humans walk around with robot servants (fig. 4.66).

2100

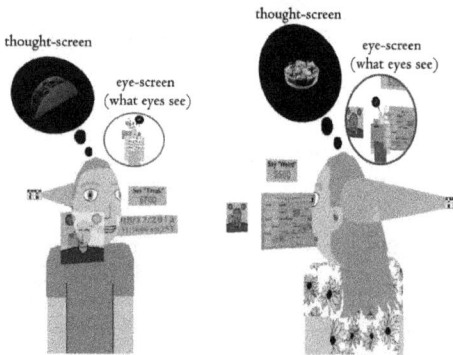

Figure 4.67. What humans communicating by thought using direct-to-brain windows may look like.

Most humans communicate only by images and sounds of thought (fig. 4.67).

Figure 4.68. Helicopters form lines of traffic above the street. Even large artificial muscle wing flapping flying vehicles are a possibility- the pterosaur "Quetzalcoatlus" shows that it's possible.

Helicopter-cars form a second line of traffic above the streets (fig. 4.68).

100 ships with humans orbit Earth.

2140

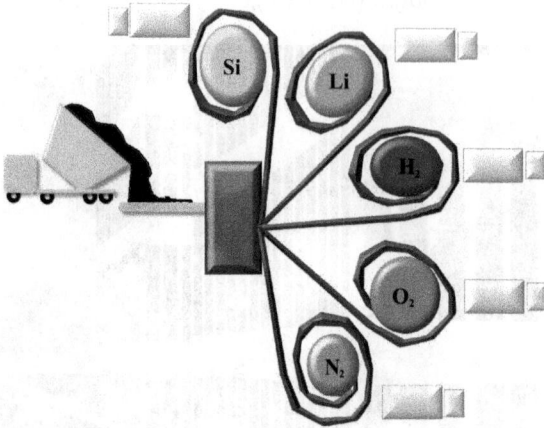

Figure 4.69. Large scale transmutation: waste goes into a particle collider, and charged atom fragments are separated by mass using an electric field.

Large scale transmutation (fig. 4.69): common atoms like Iron converted into Hydrogen and Oxygen using particle accelerators and colliders.

2220
Robots do most low-skill jobs.
1000 human-filled ships orbit earth.

2270
Humans live on Mars.

2500
End of death by aging. Growth development of a body can be made to go forward, stopped, and/or reversed by nanometer scale devices changing the order of DNA nucleotides.

2550
Humans live on Venus.

2570

Figure 4.70. Humans will probably use the masses of many ships (gravity) and thrust to move large bodies around the star system.

Humans move an asteroid (fig. 4.70).

2650
Humans create atoms from light particles.

2750

Figure 4.71. The first ships to reach a different star will be an epochal moment for humans of Earth.

Ship reaches other star (Alpha Centauri) (fig. 4.71).
First close up pictures of planets of a different star.
Living objects found around another star (bacteria
made of DNA found on planets of Alpha Centauri).

2800
Humans change the motion of a moon.

2850

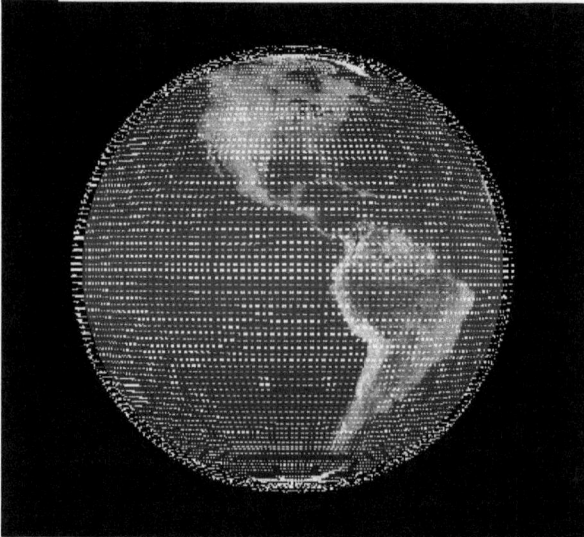

*Figure 4.72. The Earth may look like a bee-hive of ships in
the future.*

Humans change the motion of a planet (the Earth)
(fig. 4.72).

2900
Ship impacts surface of Jupiter. First image of
surface of Jupiter.

3200
Ship from Centauri reaches Earth with objects.
Humans reach a different star, Centauri.

3500
Atmosphere of Venus removed.

4000

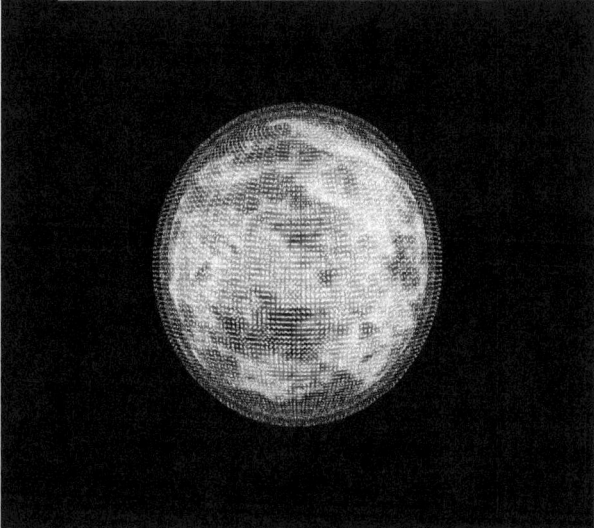

Figure 4.73. Ships consume all the atmosphere of Jupiter to reveal the massive liquid red-hot molten surface.

Atmosphere of Jupiter removed (fig. 4.73).
Humans have ships at 10 stars.

4500
Motion of all planets under control.
Humans reach center of Earth.
Humans live on Jupiter.

5100

Figure 4.74. Probably some organisms will be adapted to low gravity, and others, like us in our current form, to higher gravity.

Image of advanced life of a different star (fig. 4.74).

5500

Figure 4.75. Thousands of ships move the Sun using only thrust and gravity, without ever needing to touch it.

Motion of star controlled (fig. 4.75). Earth star moved in direction of Centauri.

6000
Humans touch advanced life of a different star.

17000
One trillion humans.

27000

Figure 4.76. How the stars humans occupy might look in 25,000 years.

Humans inhabit 100 stars and form a globular cluster of 10 stars (fig. 4.76).

47000

Figure 4.77. How the stars humans occupy might look in 45,000 years.

Humans inhabit 1000 stars and form a globular cluster of 100 stars (fig. 4.77).

65000

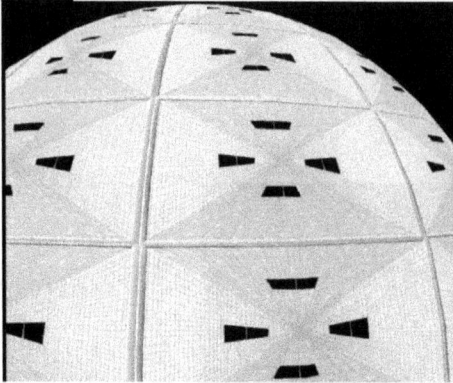

Figure 4.78. Earth is completely filled: "I live in unit 58173639203, enter in sector 25, take a left, and then a down."

Earth is completely filled with living objects (fig. 4.78).

130,000

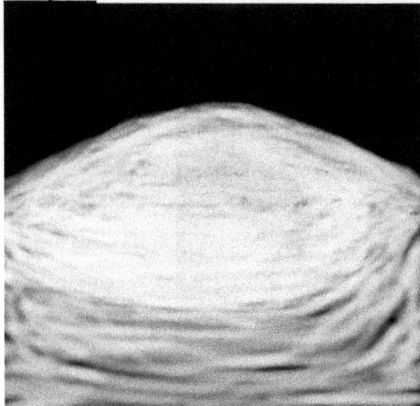

Figure 4.79. How the Sun might look in 128,000 years.

Star of Earth completely consumed by living objects (fig. 4.79).

20 billion years

Figure 4.80. How the Milky Way Galaxy may look in 20 billion years.

The Milky Way is by this time a Globular Galaxy (fig. 4.80). You would think that people in science would have told us about this likely future centuries before now, but no!

Chapter 5
More Details of D2B and Other Science Secrets

Is "neuron reading and writing" an actual science? Yes, of course it is! And so are "remote neuron reading and writing", "recording thought-sounds", and "recording thought-images". But why is there only silence about what is obviously a very important field of science? Are any governments, companies, and universities working on remote neuron reading and writing? If yes, why don't they call it "remote neuron reading and writing"? Why have we never heard of "remote neuron reading and writing" before? Could it be that there is a very strict and harshly punished restraint on talking openly about this very important and ancient science by people who perfected it centuries earlier and now trickle it out in crumbs to the many poor D2BW consumers in return for many millions of dollars?

Here are more details to help excluded people understand and defend themselves against the many vicious remote neuron reading and writing owners and consumers.

What does it actually look like to see direct-to-brain windows?

As an excluded I don't know for sure what receiving D2B looks like, but it seems pretty clear that it looks similar to a computer desktop (people have many windows open) – but the windows are in front of your eyes and are probably semi-transparent. But with D2B, windows can be anywhere in a person's view. There could be giant windows. A person could be surrounded by a cylinder of square tiled windows. Mostly people probably just have square and circle two dimensional (flat) windows. Most of the windows are probably semi-transparent, but a person could see a movie with "total immersion", that is, that the movie completely and opaquely fills their eye screen. Possibly people may view windows on their thought-screen too.

What kind of direct-to-brain windows do people get?
I don't know for sure, but it seems likely that some common D2B windows are: 1) Any window you might have on your computer, like a web browser, email, instant messages, Word Processor, Video Editor, remote desktop windows to their home computers, calculator, a clock that shows the date and time (wouldn't it be so nice to be able to quickly popup the time in front of your eyes to know if you need to hurry or have time to do some activity?), a calendar (just like many people have a paper calendar, and now electronic calendars to organize their time, and to match open times with other people in the future). The memory that stores all the events on the calendars is, presumably, at the neuron D2B provider company, which is presumably the undemocratic monopoly of AT&T and Verizon, here in the USA. There must be tight controls on the D2B data. There are obviously no videos of people and their thought-screens and thought-audio captured by microscopic cameras that have reached the public. Probably average D2B consumers cannot store D2B data in a computer in their home, the way the Internet is. Where is all the D2B data stored? 2) Video windows with a live image of your mate, kids, friends, and co-workers. These windows may stay on constantly or only appear very frequently. Since the human mind can only really focus on a few things at a time, perhaps there are just frequent updates, like instant video messages, that D2B consumers just notice and don't think much about. Clearly, many D2B consumers watch and are fixated on their enemies, judging from the large quantity of remote molestation and nasty shills that I endure. 3) Video windows of those around you: It seems likely that as a D2B consumer walks around, windows of people around them may pop up showing a quick profile of the person, and relevant videos from their past. 4) There must be back logged video-messages just like email for people that D2B consumers periodically look at. 5) Pop up "money to you"

windows. These windows must offer to pay the D2B consumer money (probably in the form of "virtual money" used to pay for more D2B services. The offers are to pay money to D2B consumers to "shill", that is, to say things, usually unpleasant things, to excluded people, for small jobs like even just to walk some where, to walk past, or walk with an excluded person, money offers for dates, and to perform pleasure-based (sex) services. Money, like a big rotating virtual 3-D "$50" sign might appear in distant 3D locations as seen by the D2B consumer, the D2B consumer then goes to that location, and the $50 sign probably disappears with a sound like "ching!", and enters the D2B consumers account; kind of like a 3D computer game, or leading ducks with bread crumbs, but with humans and money. There may be little virtual money signs on people's bodies, like on their lips for a kiss, etc. It's probably much easier to direct people's motions with little D2B virtual signs. 6) Pop up "money from you" windows: Just like there are windows that offer to pay the D2B consumers for various things, so there must be windows that offer to sell things to D2B consumers in exchange for money. This may include offers for products or services that you specifically may be currently looking to buy, and offers to pay-to-view popular videos (and videos apparently, and fortunately, with very little or no restrictions on privacy, sexual or violent content). 7) Popular videos. D2B service may cost a monthly fee, like a phone bill, but some videos may be free to view. Clearly some videos must cost money to view. Like news stories, there must be tons of videos of recent events on Earth to choose from. Probably the most watched videos are of sex and violence. Much of these interesting videos are probably funded by the wealthiest people on Earth. For example to make people have sex or do violence it must cost wealthy D2B consumers thousands of dollars. Many of the videos must be of excluded people who bit on some bad remote neuron suggestion and did something stupid that many of the D2B consumers then pay to see. You can see how the money that people pay to see popular videos funds the making of more similar videos. D2B consumers apparently see many videos from inside people's houses. Some of these videos may be free, but probably many cost money. Probably there is a constant stream of "death of person A", or "sex between A and B" video offers. Videos of murders, suicides, accidents, crimes (including sex crimes- often produced by remote neuron writing), people nude or having sex in their houses, etc. are probably all advertised and available to those who receive D2BW. Clearly D2BW consumers are not punished by the same laws the excluded public is punished by relating to privacy and pornography. In particular people probably pay to watch videos of the deaths of famous people, and famous crimes from the past (even the far past- hundreds of years ago), to know who killed who- in particular to see from the eyes and to hear the thought-audio of all involved. Probably many people are interested to see videos about crimes that involved somebody they cared about in order to know the truth about what really happened. Excluded people have to piece together the tiny fragments of evidence available to the public, but D2B consumers probably have most questions answered about who is responsible for any given crime. Many people probably want to see videos of their ancestors and their thoughts as far back as the secret videos go. There is probably a lot of watching the "most beautiful" and "most interesting" people. Many D2B consumers are clearly interested in seeing those excluded who know about D2B (all 3 of us) and in particular that are publicly telling the truth about D2B. In terms of what videos D2B consumers see, there is a parallel, although extremely less restricted, with the Internet. So, for example, if you are a person who dreams of moving to the Moon or Mars, you probably want the latest videos about the forefront of that effort, etc. 8) Possibly some kind of

Neuron company administrator people popup windows: It may be that the neuron administrators popup to communicate, or possibly even just text messages to the eyes of D2B consumers- many times apparently warnings, teleprompting, or perhaps coaching what to say out loud and/or what to do. 9) Remote neuron drawings that are made to convey some message. Many times, even as a D2B excluded, I get little tiny visual (and certainly many annoying audio) neuron writings. One example is how a little white dot goes down, and this may be sent to make the receiver think that something they have thought or done has lowered their popularity or is receiving disapproval among D2B owners and/or consumers.

The "Eye-Screen"
Unlike windows on your computer, the windows of D2B are probably mostly of people, and all people probably have two circular or oval screens, over, or around their head (like Mickey Mouse ears): their "eye screen" and their "thought screen". You may hear D2B consumers talking about seeing "eyes". They are talking about seeing what people's eyes see. When you can remotely read what the eye neurons see, people become like walking cameras. The eye screen shows what the brain's eyes see. If you are looking directly in the eyes of any organism with eyes, seeing their eye screen is like looking in a mirror because you see what they see, which is you. We tend to view the image our eyes see with a lot of depth, but the image in our eyes is a two dimensional image that can be copied and sent around like any JPEG image. There is no need to take a photo of something anymore, because we only need to look at it. But, the current problem is that none of our past or present eye images are available for us to access. So we still need to capture images with our own cameras if we ever want to see the image again.

You can imagine that images sent to the eye screen could be used to trick a person into thinking that some object is there when, in reality, the object is only drawn there. For example, have you ever thought that you saw something, and then looked again and it wasn't there? That probably was a millisecond neuron writing to your eye neurons. Another classic is drawing (mapping) the image of somebody you know over the face of somebody you don't know for a millisecond. I have no idea why those with neuron writers do this. I have experienced this numerous times, and even my Mom recollected this phenomenon which happened to her while in college - in the 1950s.

The "Thought-Screen"

Just like there is an "eye-screen" that shows what a brain is looking at, so there is a "thought-screen" that shows any image a brain is thinking of. It seems likely that, most of the time, a thought screen is simply black, which indicates that the owner of the brain is not visualizing anything in their mind at that instant. Think of an image of some food that you like to eat. That image that you just thought of appears on your thought screen and can be captured and sent to other people as an image (.jpg) file. For many people, it may be that only occasionally and briefly, an image will flash on their thought-screen which shows an image that they are thinking of or remembering. For example, when an image of food appears, it most often means that the person is thinking about what they want to eat. An image of a place the eyes saw once may appear, for example, the image of the view in a grocery store aisle, which may indicate that they are deciding if they may go to that location sometime in the future. An image of a person shaking hands with somebody they see means that they are thinking "should I or will I shake this person's hand?". It may be that D2B consumers have many more images on their thought-screens, knowing that such images can be seen, and have trained themselves to draw images there. A thought-image can convey a person's instant fears or impressions – many times in a comical way- for example, a person asks another person if they want to buy a book, and the receiver of the offer has a thought image of themselves sleeping with a big "zzzzz" over their head. So the thought screen of a D2B consumer may be much more animated and brightly filled than the thought-screen of an excluded, but I don't know for sure.

One really amazing aspect of both the eye and thought screen is that even other species have an eye and thought screen. There is a distinction between just seeing something, and being able to "redraw", or "remember" the image the eye captured in some array or layer of neurons. The ability to draw or remember images on a thought-screen probably is found in all species back to the evolution of fish. Ctenophores (TΘNΛFORZ[122]) and Cnidarians (N⊦DΛRΣⱢNZ) (sea pens, sea anemones, jelly fish, and corals) are the most primitive multi-cellular organisms to have neurons. Many jellyfish and other invertebrates like worms definitely have an eye screen, and also may have a primitive thought-screen which occasionally shows images. I think certainly many birds and mammals and probably even reptiles, amphibians and fish can produce images written on their thought screen. Which species is the most primitive that can draw to a thought-screen?

"Ears", "Hearing", "Ear-neurons", "Ear microphone"

Not only is everybody a "walking camera" because their eyes capture images, and a "walking billboard" with a screen that shows their thought-images and images externally sent or activated there, but we are also walking microphones that record sounds around us. In addition to the visual neurons of the

---

[122] See my one-letter-equals-one-sound phonetic alphabet to hear how each letter sounds at http://tedhuntington.org/fonik.htm.

"eye" and visual "thought" screen of the brain, there are auditory neurons that can be read from and written to. "Ear neurons": *Reading* from neurons of the ear allows people to hear what the body hears externally, for example, a song from a speaker, a bird chirping, a car passing, etc. – but all done remotely, through invisible frequencies of light particles, without having to touch the brain. *Writing* to the ear neurons can make the owner of the brain think that they hear external sounds, like a song from a speaker, a bird chirping, a car passing, etc. again, all done remotely using invisible frequencies of light particles and many tiny floating and stationary devices inside and outside of the body.

"Thought-Audio", "Thought-Sounds", "Inner-Ear"

Just like there are auditory neurons that record external sounds, so there are auditory neurons that record the internal sounds of thought. To *read* from these neurons allows a person to hear the sounds a person is thinking of. For example, think of a song. Then think about being able to record that song in your thoughts and send it to other people just like an mp3 sound file, again all done invisibly with low frequency light particles. To *write* to thought-audio neurons produces sounds in the body's thought-audio. For example, sending the sound data (like from a .wav sound file) of some song to a person's thought-audio neurons, makes them hear the song, not externally, as if the sound is being playing from a speaker, but *internally*, as if the sound is located inside their head. It is, of course, possible, to play audio of the voice of the brain's owner, which might trick a person into thinking that they themselves actually thought something that was externally written to their thought-audio. Even activating neurons that form internal thought-audio sounds might be possible, which really make a person feel that they truly created the thought-audio themselves.

The hearing and thought-sounds of the other species can definitely be read from and written to.

Imagine how interesting that would be to those of us who have never heard thought-audio from a different species. For example, hearing a horse play back a song from the radio, or recalling a human voice. Which species were the first to have thought-sounds?

Other sensory neurons

Presumably, any neuron, inside anybody, can be read from and written to, and there are many complex neurons. A brain can be made to feel sad, to laugh, to cry, to remember some image, sound, smell, taste or physical sensation from long ago in the past- presumably just by making that particular neuron fire, or perhaps by making that neuron have the highest relative electric potential (voltage) of all the neurons in the cerebrum.

In addition, neuron writing is used to make you "aware" of some part of your body, that is, to make some small part of your body the focus of your current thoughts. Neuron writing can make you feel pain anywhere in your body, can make you feel like you are getting an enema, that your heart is pounding, and the feeling that you need to itch some part of your body. Beyond remote neuron writing, it seems likely that there are microscopic particle devices that can simply make a person feel an itch on their skin, without neuron writing, but just by sending light particles at the skin. But activating a neuron organelle is probably the standard and most commonly used method. One response to the constant remote molestation is to fire a mental laser back at the particle molester in your mind.

Motor neurons

Because there are neurons that are connected to muscles, which are called "motor" neurons, remote neuron writing can be used to contract any muscle of any body that has muscles – and all multicellular animals except sponges have muscles. Many people do not realize how important muscles are to live. There are many muscles that can be remotely contracted to seriously hurt or kill a person very quickly, for example simply remotely holding the lung muscles, contracting arm muscles when driving a vehicle, or leg muscles when near a steep ledge.

Figure 5.1. An office meeting where 4 people are included and 2 people are excluded. Included people probably have a D2B window of each other person in the meeting. Notice how those who are not getting D2B look "sterile", and disconnected. See d2bw.org for a closer look at the above scene in a video.

So, direct-to-brain consumers probably have a number of windows in front of their eyes. Probably people mostly have a window constantly open for each person that they care about: their mate, family, friends, coworkers, and neighbors. Each window probably shows one circle (or oval, or rectangle) with a live image of the person from the front (the person's face), another circle with their eye image, and another circle with their thought-screen, in addition to being able to see the D2B windows that the person may be looking at. Perhaps they only just get updates of relevant videos without having a constant live image.

D2B consumers don't need to look at you to see you, but you need to look at them to see them. It seems likely that when a D2B consumer walks by anybody, a D2B window must pop open with: 1) a live image from the front of the person walking by (their face), 2) a window with what that person's eye's see, 3) a window with their thought screen, and their thought-audio starts to play in the thought-audio of the D2B consumer. So an excluded has to look to see what a person looks like, while a D2B consumer can stealthily and sneakily watch a person around them for a long time without having to look at them. Not only does the D2B consumer get a live image from dust sized devices around and inside the excluded person, but probably gets a window with "highlights" from your life (your most memorable moments), and/or a quick profile about your beliefs, whether you are married, have kids, etc. We all simply want to know about those around us to protect ourselves, to look for friends and mates, but also to learn more and simply to "people watch". You can see how, obviously, this natural curiosity, fear of the unknown, and desire to know the actual truth is what drove and still drives the creation of microscopic devices that fly around and send back images from inside houses and of thoughts. There may be a quick set of ratings about the person, like how many people have rated their physical and intellectual qualities. One important note is that there may be some kind of "privacy" custom where D2B consumers don't get to see other D2B consumers in their houses or their thoughts, but you can be sure that no such custom or courtesy exist for those of us who are excluded; even the worst D2B consumers can probably see inside the house and head of any excluded person they want to. D2B consumers know that they can just look at our thought screens and thought audio and inside our houses without paying us a cent, whether we want them to or not. This includes even writing to excluded minds, because unlike a D2B consumer, the payer and the writer taking the money to neuron write know an excluded cannot see them or put up a defense in any way – it's like beating up and taking candy from a baby or defenseless child. And then there are always a million flimsy excuses as to why this kind of barbaric segregation and inequality is ok.

Think of the worst things you have ever done, or the most sexual, and/or most embarrassing or most violent things you have ever done. Probably they were all remotely neuron written suggestions, as crazy as that sounds. Those are probably the first and most popular videos of you that the D2B consumers see. We all have them, so you are definitely not alone – some are much more violent or sexually embarrassing than others. You went out in the nude, you lied, you stole, you assaulted somebody, you wore opposite gender clothes, you did something sexually inappropriate, unusual, or embarrassing, etc. D2B consumers have far less of them, because, of course, they know to reject remotely written suggestive images and sounds, and that anything they do will be used as a weapon against them for the rest of their life, and may greatly restrict their monetary and mating possibilities. The D2B owners send constant sexual, suicidal, violent, unhelpful suggestions in many different forms, very fast thought-images, thought-sounds, etc., and excluded people have no idea that somebody is sending them some terrible suggestion, they simply repeat it.

Because people can hear your thought-audio, thinking thought-audio is like talking. It's very tough to remember this fact, in particular when you don't hear the thought-audio of other people. D2B consumers may have learned to stop all thought-audio, for example during meetings (see fig 5.1), because producing thought-audio during a meeting is like rudely talking out of turn. But excluded don't know about this social practice, and so listening to the thought-audio of meetings and group gatherings must be kind of funny, because one or two voices-the thought-audio voices of the excluded people-continue on while all the thought audio of the D2B consumers is quiet. But also it must feel disgusting, to leave other humans in such a 1200s kind of state.

Walking down a street, passing people, you have to realize that a little window pops up of you, your eye screen and your thought screen, and your thought-audio starts playing in the D2B consumer's thought-audio, which they can control the volume of, and completely mute if wanted. But probably, like talking, most people probably like to hear all the thought-audio around them. So walking by people on a street, you have to realize that your thought-audio is like you are talking out loud – the D2B consumers hear it all. They also know what you are looking at, in particular, if you are looking at their butt, breasts, etc. because they recognize the picture of their butt in the window showing what your eyes see.

If we see an image in our mind, for example the image of a funny face, we decide quickly if we want to echo, or act-out, that is, move our muscles to imitate that same funny face. Neuron writing is many times just to plant a suggestion, and to make the context of thought some particular topic. For example, they write an image or a smell of a pizza for 1 millisecond, just long enough so that you notice, and your mind then decides if you want to eat pizza, making a yes, no, or no decision. Neuron writing is the same thing as seeing an ad, but these ads are sent directly to your brain without you seeing the sender, without the receiver knowing even that such technology exists, and so they are very persuasive. Once a person realizes that humans are sending images, sounds, and smells they can step back and resist an instant acceptance – or at least have some skepticism.

Is that song in your head you just happen to find yourself humming sent there by people with bad intentions? Probably. One surprising truth of remote neuron writing on excluded people is that many popular songs may have been partially secretly remotely written by violent wealthy people who buy remote neuron writing to influence a large audience. There was a funny skit on MTV where Randy of the

Redwoods is reading a newspaper story out loud about how a killer is coming after him, while a man approaches him and appears to be following the story line Randy is reading, and so then Randy ad libs by changing the story to read that the supposed killer then goes far away, prompting the approaching man to move away. This is a good analogy for the way neuron writing is used on excluded people to try and get them killed. One example is a lyric in John Lennon's "Out The Blue" which is "...you came to me, and blew away life's misery...", could this have been remotely written onto Lennon's brain without his knowledge, that is, without Lennon realizing that such a message might be misinterpreted to mean "blew away" with a gun? I myself hear many songs in my head that, me repeating and whistling out loud, would not be to my best advantage – songs by notorious homosexual people, for example. I am completely a supporter of all bisexual people, but this kind of neuron writing is used to try to falsely label or influence excluded- to make an excluded person appear to be homosexual. Terrible D2B owners and consumers constantly try to explain the apparent lack of success excluded people have finding mates as due to homosexuality. But in reality, the lack of success at finding mates is obviously because of the neuron apartheid- excluded people simply cannot see potential mates and their thoughts. Much of the remote neuron writing is designed to divide like-minded decent people by focusing on non-violent issues like "gay?", "perv?", and "insane?" instead of "violent?". The argument given by the neuron consumer buyers and neuron owners for such suicidal, homicidal and unpleasant neuron rape writings of humans who have never even heard of remote neuron writing is that it is simply "free speech".

Classic remote neuron writings
There must be some often used traditional remote neuron writings that are centuries old and have a long, but publicly secret, history.

Motor neurons
Lung muscles
Of course, a person's lung can be held so that they are suffocated – and because the particle frequencies are invisible, and the remote murderer cannot be seen, there is nothing any coroner can say other than- "here is a perfectly healthy person that just died, without any evidence of a cause of death". Holding the lung muscles may be one of the most popular forms of remote murder, and probably more than 1 million people have been killed by somebody, or by a group of people remotely keeping the lung muscle of the victim contracted or uncontracted until death by suffocation. Many times a lung muscle may be remotely and invisibly held when a person has suffered some kind of damage, or has cancer and is in a hospital, or uses recreational or prescription drugs, because there is much less suspicion when they suddenly die. It may sound crazy, but illness, cancer, and drug addiction are helpful excuses to explain why a person mysteriously and suddenly dies. Possibly among those killed by this method of suffocation are, not only my Mom, but many famous popular liberal people like: Michael Jackson, Elvis Presley, Jim Morrison, Jimi Hendrix, Janis Joplin, John Belushi, and Marilyn Monroe.

Arm muscles
When a person is driving, the remote neuron writer contracts an arm muscle to make the victim drive into an accident. Could this be what happened with Teddy Kennedy, and a person speaking out about 9/11, Danny Jowenko?
Another classic motor neuron writing is moving your arm and/or hand to make you cut yourself.

When you are cutting a vegetable, remote neuron writing can move your hand slightly to cut your finger, and the same for peeling a vegetable, or to burn your hand on a hot rack in an oven, or to get stabbed by a knife in a dishwasher, etc. I have even had my hand cut with my own fingernail. Computers analyze sharp objects, and can make use of them to hurt you in their 3D rendered model of you and your surroundings. One piece of advice is: don't hold onto sharp objects like knives for a long time, carefully and consciously make each cut, and when using sharp objects, like drills and machine tools, get in and get out, because advanced computer programs can easily remotely and involuntarily move one of your muscles a few millimeters to cut your body.

Making you cut yourself during shaving is common; by simply applying a little more muscle contraction than is required, or moving your arm just a millimeter too far to cut your nose, etc. There is really no way of protecting yourself, because the remote particle technology is obviously more powerful than we are, but you may be able to reduce the remote cutting by taking extra precaution when using and working around sharp objects.

Leg muscles

A quick muscle contraction can send a person off a roof, bridge, cliff, out the window of a tall building, into another car, etc. We can be optimistic that self-driving cars will remove much of these remotely caused accidents, but even remote transistor writing will probably be a problem. The best answer is to identify and stop the people behind all the violence-causing remote writing.

Finger muscles

Making a finger twitch is popular. One use is to deprive a person of sleep. Just when the victim falls asleep, their finger is made to move, and this wakes them up. Once awake they cannot go back to sleep. Sometimes a right middle or index finger will be twitched to indicate that the victim is not "right-wing". A finger might be twitched to make an excluded person think that they have a health problem (like carpel tunnel syndrome), or feel less confident because they mistakenly believe that they are getting older and their fingers or hand is shaking like an older person.

Sometimes tiny cuts can appear on your body – mostly on your fingers – and then your right middle or index finger. Perhaps these cuts may be done with particle devices that do not communicate with neuron organelle devices, but probably the neuron organelle devices can be made to move around or cause similar damage within the body.

Eyelid muscles
The tiny muscles that control the eyelid are sometimes made to twitch, and this is supposed to imply that you have a mental disorder.

Eye muscles
Moving your eye muscles is one of the trickiest muscle moves, because it's hard to believe that even where you look, where you point your "camera" (your eyes), can be remotely controlled. Eyes are moved to make you look at something: a knife, a butt, an ad, a license plate, etc. Sometimes there might be a two part molestation, like your eyes are made to look at a knife as you walk by it, and around the same time, an image of a friend is flashed on your thought-screen. You can easily see the quality of the people that run remote neuron writing.

Vocal muscles:
There is also the so-called "Freudian slip", which is when your vocal motor neurons are written on to make a verbal mistake; for example to say something that was not what you intended to say, but that was somewhat comical, although usually from a low-brow perspective. Probably 99% of so-called "Freudian slips" are remotely written and are

evidence that very fast computers not only understand the perspective of the person paying for the neuron writing, but also what makes people laugh. Because the computers are faster than the humans, perhaps a person just buys some verbal neuron writing without knowing what specifically will be done, and the computer just works with what the victim is saying at the time.

Sensory neurons
Touch neurons

Itches: For example, you have to scratch your face and in the process you make a right or left arm salute. The "left" "right" analogy is very popular to the wealthy who constantly do neuron molestation, probably because a small group of wealthy violent criminals wants to hide behind a large group that identifies with being "right", but also to focus poor people against each other with constant "right versus left" fighting instead of at them. "Right vs. left" is a major theme for the evil side of the neuron. Like the "Reich" (right) of the Nazi's, perhaps one of the biggest convincing divider of people is this concept of "right" and "left".   The murderers, assaulters, remote particle molesters, and liars all band together under a banner of "right" and all those that refuse to obey or ever disagree are, of course, the "left". "Blue versus Red" has also been a popular method to divide people, but "left and right" is apparently more popular for the neuron. Directions like "middle", "up" and "down", "forward" and "backward" are not as popular. As an analogy with the human body, a more logical view is that both left and right sides should be equally utilized, exercised, trained, and developed- anything else results in a body that is not symmetrical. Beyond that, how violent and dishonest a person is, is a much more useful, clear, and important way to distinguish between people.

Ear neurons

Remotely writing to ear neurons is perhaps the number one most popular form of remote neuron writing. Many times the audio sent to a person's ear is some kind of verbal abuse. Many times a gruff voice in your head may say "perv!", "vert!", or "nut!". Other times a female voice may say similar words, probably because the 90% male owners of neuron writing want to somehow shift blame and anger to the mostly powerless women of Earth. A bizarre part of the D2B is how they all secretly watch people's thought screen while the unknowing excluded people are showering, but somehow the excluded are the "perverts". Writing thought audio like this may be to reinforce the myth that somehow those who own and receive D2B are sexually and mentally normal, but that those excluded are excluded because they can't control their behavior and their sexual desires, or have unusual behavior and unusual sexual desires. Also the goal is to change the focus to an antipleasure and sanity context and away from exposing the murderers, going public with D2B, and other more important and honest topics. The worst part is that the hypocrites that abuse excluded people, send all kinds of sexual suggestions – to go nude in public, to put a banana in your butt, to put a door knob in your butt, to wear the opposite gender clothes, to sexually touch people- in particular people under the age of 18, etc. Then the excluded bites on the suggestion, and may get in serious trouble, lose many friends, their job, and be viewed as a fool, but all those distant neuron writers are never even seen, let alone blamed or punished as being probably the main reason that any kind of unusual behavior occurred in the first place.

Other classic thought audio writings are "shutup!" and "rat!". Ironically, the neuron writers never shut up with their constant thought-audio molestations and big lies, and while you can't see one image of the offending writers, they have unlimited access and freely share (or "rat") all your information including images of you in your room, and your every thought-image and sound.

Some of the verbal abuse in your ear is used to reinforce a myth that the excluded is dangerous, like a woman's voice saying "leave me alone!". But, of course, any thought of leaving the excluded person alone and not writing to their ear neurons is not of concern. Sometimes the audio "leave me alone" is flashed with an image to associate the audio with a person. An excluded might become angry at such an accusation and wrongly blame the owner of the voice. It seems almost certain that the voices and images of people sent with D2B are most often "identify theft" and impersonation. Synthesizing a voice and sending audio samples is a simple thing to do, and given all the violence done by the neuron owners (9/11, the Kennedy murders, all the galvanizations, etc.), impersonating other people by sending sounds of their voice to people's thought audio is minor in comparison. Probably many excluded people instantly believe that their friend really did say the terrible audio in their mind, even though it never happened; the audio and images were synthesized and remotely written to their thought-audio and thought-screen.

The technology has evolved over many years, so sounds like the creak of a chair, or door can be "shaped" in the mind to sound like a voice. A distant voice can convey some paid-for remote neuron message of hate. The indistinguishable sound of people talking in a distance can be shaped or similar sounds overlaid to make a person recognize some audio phrase that is never actually said.

Eye neurons

On excluded people, eye neurons are more rarely written on, and when they are written on, it is usually extremely fast. But when a person is sleeping, there may be a lot of remote writing to the eye neurons. In fact, probably most of the remote writing to eye (and thought-screen) neurons is done while an excluded person is sleeping, or when they are just coming out of deep sleep. The science fiction writer Hugo Gernsbach talked about people of the future remotely learning while sleeping by using remote neuron writing way back in 1911, and there may be very wonderful dream movies that people can buy to see while they sleep (and interactively be a part of – as if you are in the movie), but in my experience most of the movies sent to dreams are terrible. I've had dreams where my friend had their head cut off, where my Mom jumped off a balcony to the pavement far below and died, where a friend was chopped with a hatchet, where I was chased, where I was shot at by a person with a handgun, in one my Mom was shooting a gun at me, where I was trapped in a hall that kept getting smaller and smaller, where a door opened when I was alone and a person suddenly attacked me from the darkness behind the opened door, where many insects like spiders were chasing me and crawling on me – many times the big insects are stuck on the dreamer's body- like on your arm and you blow and swat at them (while still dreaming), but they don't get pushed off– those are just the ones that come to mind. Because the dreams are in front of your eyes, it's exactly like these terrible events are really happening around you, and I'm sure can cause a heart attack or stroke just from shock and surprise of what people experience in these scary and violent neuron written on dreams. I often awake, from bad dreams, and knowing that they are sent there rapidly fire back at the senders in my mind, but it's probably useless. Probably the evil half of the Neuron sells these videos to like-minded consumers who then laugh to see their enemies tortured in their dreams. You may even recognize the dreams I describe above, because they probably get replayed on the eyes and ears of many poor victims. On the opposite side, very rarely, there have been a few sexual dreams that were nice, like where I got to kiss a pretty woman or got oral sex, and then there was a dream where there was a complete musical and the chorus of the song that many of the people sang and danced to, which I still remember even now, was "when you feel some motion!". But I never get any movies where we go to the other planets, the other moons, the other stars, or meet living objects of other planets – just most often the lowest brow, most terrible dreams. It's probably no wonder that we receive stupid and scary neuron written dreams, when so many people bizarrely embrace "horror" and other violent movies, and the idiocy of religions. All this time we could be seeing the thought-screens and dreams of people and the other species, but the wicked D2B owners have kept it all for themselves.

When, on the extremely rare occasion that an image is sent to an excluded thought-screen, it is almost always an unwanted image, for example an image of a person who you don't like or don't get along with, or even simply don't want to be reminded of.

Other neuron written images you might receive:

Images of food. Probably obesity and being excluded are closely linked, because people constantly send images of (and impulses for) food – but so quickly that a person who is not aware of D2B doesn't realize that the food images (and impulses) are being sent to their brain remotely by another human. Then, of course, a person decides whether to eat the food in the image. The chances of an excluded person blindly following the suggestion are probably high. Many restaurant chains and food companies may send images of their products because they know from experience that they get more business by planting the suggestion of their food item on the thought-screen of excluded people. Those companies who don't neuron write ads may lose money or fail altogether. Making an excluded person fat also lowers the popularity, and therefore the influence of the excluded person. But more sinisterly, remotely neuron written food suggestions to make excluded people get fatter are used not only to increase an excluded person's risk of health problems, but also to lower their chance of getting sex, and therefore of reproducing, and so are used by wealthy people to try to lessen the influence and ultimately to exterminate their enemies (political, religious, philosophical, etc. opposites) among those who are excluded.

Images to make you do something sexually inappropriate. When the goal is to lower the popularity and chance of reproduction of a smart, honest, talented, educated, good-looking and/or heroic excluded person, making them bite on inappropriate sexual suggestions tops the list of neuron writes. Planting an image of an underage friend while the excluded masturbates so the excluded person then includes images of the underage friend into their masturbation thought-images is probably one classic technique. Thought-images that people have during masturbation (and sex) probably generate a lot of money for AT&T and the neuron owners. Excluded people tend to be more open minded about sexuality, in particular since they don't realize that many people are watching them and their thoughts. Much of the key to getting excluded to do sexually inappropriate things, is that the excluded simply has no idea whatsoever that they are always on "candid camera". This idea is unheard of; they can't possibly realize that microscopic cameras and neuron reading and writing devices are a very common part of life on Earth, no matter where they are.

Images to make you steal. Another way to lower the popularity of popular excluded people is to get them to steal. It's amazing how easy it is to get a decent excluded person to steal when they don't think that anybody can see them.

Other images. I have seen images of a tattoo on my body for a millisecond when I look in the mirror, and thought "how would I look with a tattoo?" and "Would I ever get a tattoo?". It prompts the question:

"Get a tattoo Y/N?" in the thought-audio of excluded people. I have seen piercing jewelry (recently one in my navel). Probably at least 20% of excluded can be made to get a tattoo or piercing using this method. Although some people find tattoo's attractive, creative, and a turn on, probably for most people a tattoo (like obesity, violence, and other acted-on D2B suggestions) can significantly lower a person's reproductive and economic value. The senders of the D2B image know this (but most excluded don't), and this method of "lowering the value" of political enemies (usually those for democracy and majority rule), and mostly excluded people, is an industry that many excluded are the victims of.

As an excluded who has never really seen a thought or eye screen – it's tough to see how people act and think, but for those people who have been watching eye and thought screens for centuries, human behavior must appear to be very simple. So I can only kind of give you a basic view – videos that show thought-screens would much more quickly and effectively show you how people think and how they are tricked with remotely sent images and sounds.

Memory neurons

One very subtle form of remote neuron writing is to activate memories that then relate the message the buyer of the D2B writing wants, for example, to activate neurons of unpleasant memories – or memories of events where the evil side "won" – like a successful shill from a person who normally resists the money to shill. Remotely activating an earlier memory may be less expensive then writing a new thought-audio or image, but also, may feel less artificial and less externally produced to the victim since the memory is one they have felt or seen before.

The neuron "organelle"

For a long time I thought that remote neuron reading and writing was done just with light particles in x-ray frequencies directly to the neuron, and maybe that is true, but it seems likely, as crazy as this sounds, that micrometer and/or nanometer scale devices, like some kind of RFID devices, which are very small – smaller even than dust, may function as the first human-made cell organelles (like mitochondria or plastids but made by humans). Another theory is that particle devices, probably microscopic, send light particles of invisible frequencies right into the eyes and ears to draw pictures and create artificial sounds without any intracellular devices. Such external devices could also do remote neuron writing directly on the neurons of the motor and pain (nociceptor) neurons by penetrating the outer skin (which of course we all know high frequencies of light particles can do). I doubt such a device could do remote neuron reading very well. Possibly, they could see eyes and hear thought from the sub-visible frequencies of light emitted from the brain, but since there are many layers of neurons emitting light particles, it might be difficult to isolate particles emitting from just neurons responsible for smell, for example. Maybe this kind of technology was developed initially, before the "ah much better!" moment when a neuron intracellular electronic device that assists in remotely communicating with specific individual neurons for reading and writing was invented. To my knowledge, nobody has ever published a method showing how to remotely activate neurons using only light particles, but there is very little published on those kinds of experiments.   Without an intracellular device, I think it would be very difficult to read from or write to individual neurons, but such technology might be accurate enough to remotely make a muscle contract, cause sounds to be heard, to make a person feel an itch or smell a smell, and to make an image appear in the eyes.

It seems likely that to remotely read and write to individual neurons, in particular those deep inside the brain (like activating a past memory), there needs to be intracellular devices. These tiny dust-sized devices can enter a body through the lung, enter the blood vessels, and eventually enter various cells. These devices might use a network (MAC) address, be remotely steered and moved with tiny motors, transmit the electric potential of the neuron or any other cell that they are in, and be able to change that electric potential – to make a neuron fire. A neuron is a lot like a transistor, or electronic switch that is either on or off. One of the best pieces of evidence for this theory is the scene from the 1967 movie "The President's Analyst" where the phone company representative explains how a chip enters the body and is used to make phone calls using only thought. Since remote neuron writing using only light particles has, so far, never been proven publicly, it may be that some kind of device needs to be in a neuron to make it fire. In addition, having an electronic "cell organelle" device with a numerical address, removes the problem of directing a signal to some precise neuron location; instead a stream of light particles in an invisible frequency, that contain an encoded address, can be sent in the general direction of a person's head, and only the specific device with the correct address will respond to the signal, firing the specific neuron, or sending back the neuron's value. Another strong piece of evidence in support of the theory of very small chips functioning as cell organelles is that this is the basis of some of Galvani's famous 1780s experiments. Many people don't even know that Luigi Galvani did remote neuron writing by 1791 using a spark machine and holding a scalpel directly on a frog nerve. Imagine the process of cranking the spark generator as still the same, but replace the scalpel with a much smaller piece of metal. The metal amplifies the light particles moving from the spark, raising the electric potential of the nerve cell causing it to fire. What is shocking is that, not until November 5, 1959, with the remote control pacemaker, was there one published report, to my knowledge, of this simple extension or optimization of Galvani's spark and scalpel experiment of making a neuron fire remotely, but with modern technology or even a much smaller cell-sized piece of metal. This is not to say that remote neuron reading and/or writing cannot be done without such a device, certainly remote neuron reading can (as Kamitani, et al showed in 2008). Perhaps remote neuron reading and writing without an intracellular "organelle" device was how RNRAW was first uncovered way back in 1200 or whenever it was first discovered. The history is completely secret, but clearly there must have been transitions like: the first "micrometer", and first "nanometer" scale device, etc.

It's hard to believe that there could be microscopic communication devices in our bodies. But for comparison, look in sunlight, for example in your car, or in the light under a lamp against a dark background, you should see thousands and thousands of tiny pieces of dust and fibers. Those pieces of dust, that we normally don't see, are perhaps 1 mm in size. By today's electronic standards, 1 mm in size is actually very large. The Hitachi "uChip" RFID chip is micrometers in size. So looking at those thousands of pieces of dust, it is easy to imagine that we inhale dust all the time. And that those tiny dust fibers enter our lung and some of the smaller objects then enter our blood stream. Even some microscopic diatoms float in the air and enter our lungs. These tiny devices can easily be powered by ambient light particles, even of lower frequencies like radio and heat, for example that exist inside the body, or even from the natural chemistry of molecules inside a body.

Couldn't we find such a device if they are in every neuron?

Possibly somebody could, in particular somebody who dissects animals that has access to an electron microscope, but probably most of the people in those fields are either D2B consumers and sworn to secrecy, or excluded and don't have any idea that such devices could be there. But even if there was an excluded that wanted to, there is one really terrible but interesting possibility: that with remote neuron reading and writing, any object, in particular small objects can be made "invisible" to the owner of the brain simply by overwriting the signal to the neurons. This may be a basic component of remote neuron reading and writing – to basically "erase" any floating remote neuron device that reflects light into the eye of a person. The neuron readers can analyze the image captured by the human eyes in milliseconds, long before the human brain can, and so it can then cover over any device – as crazy as that sounds. It's similar to the classic question of – is what we are experiencing actually there or is it all being remotely written to our neurons, and we are actually in some extremely different place? Alternatively, could we be made to live in a permanent alternate reality simply by remote neuron writing?

Remote particle molestation and assaulting
There are apparently "career" remote particle molesters who do nothing but pay and get paid money to remotely move people's muscles and write annoying neuron writings – in particular on to the brains and bodies of excluded people. These scummy D2B writers use remote neuron writing to make people itch, to deprive them of sleep, to make them stub a toe, to spill a drink, to drop some object (most usually soap), to make a gesture (most often a right or left "salute"), to cut them, and similar such annoying remote neuron writing.

The arrogance of many D2BW owners and consumers is unbelievable. Not only do they get to see inside houses, but they also get to routinely see and hear everybody's thoughts- they have unimaginable advantages over the excluded people of Earth- but that isn't enough for them- they have to remotely molest, assault and murder too!

It's unbelievable that such idiotic, annoying and petty people could be in such high positions of power and so involved with remote neuron writing – the "neuron" is absolutely upside down without any doubt and constant particle remote molestation is one of the best pieces of evidence of that. The tiny good that does come out of it is that a very few excluded people may start to wonder about remote particle devices when they feel an itch, or one of their muscles is moved. Sadly, most people probably visit a doctor and think that they have some kind of disorder like carpel tunnel syndrome. As I said in the introduction you almost certainly are a victim of remote neuron writing on many occasions and did not even realize it.

Speaking for myself, and no doubt, many other people, I am constantly being remotely molested (made to itch my face, sent some "prank call" thought audio put-down, my muscle made to stub my toe, made to say something wrong or to forget something, getting accidental cuts, etc.). One possibility is that some people are remotely molested unless they remotely molest somebody else (for example an excluded person), and so you can see, how, in the tough, survival-of-the-fittest environment we all live in, how there may be a constant remote particle "tug-of-war", "shoving match", "kill or be killed" kind of struggle going on. It may be that somewhere there is a much more violent and painful remote particle tug of war going on, which is reduced to just some smaller remote molestation by the time it reaches us.

Some remote molestation is very subtle. For example, when a person is going to speak publicly, remote neuron writing may be used to make their heart muscle pump strongly, or to move their vocal muscles to quiver a little, which may cause a loss of confidence. The best response is the usual "fire back in your mind". Then you find that, many times, as if by magic, your heart stops pounding, and the nervous feeling suddenly goes away. Many times if you receive a positive, supportive, thought-audio, you are instantly hit with terrible remote neuron writing- a common example is your thought-audio voice commanding the positive thought-audio senders to kill themselves, or a put-down to them, and other similar terrible neuron writings. I usually immediately fire back on the remote molesters in my mind, and often try to counter with positive and supportive thought-audio to the positive thought-audio senders.

The voice and images in our thoughts are our most intimate, influential, and trusted advisor. For that reason if a technology was invented that can remotely and invisibly write sounds and images to a person's thoughts, you can see how extremely irresponsible keeping it a secret would be. In particular, people are starting to realize that, not only is that voice (and that image) in our head not the intimate, influential, most-trusted advisor we thought it was, but that most of the sounds and images in our thoughts are being sent there by our enemies- people with a violent, dishonest, and criminal history. Those who own, control, and participate in remote neuron reading and writing should show even the tiniest bit of decency by at least *showing* and *telling* the world about RNRAW in the interest of protecting all those inhabitants of Earth that are being unknowingly abused by it.

Other subtle remote writings, are the negative feeling or impression you may instantly get about your friend or another person you are talking with or walking by. Remote neuron writing is used to create and amplify negatives, criticisms, and flaws, and to avoid positive thoughts, like compliments. Many times that voice in our head, or even a very subtle negative feeling in our mind, suggests that our friend isn't cool, or is weird, is unfashionable, is not very smart, has ugly physical features, is mean, has flaws, etc. Some of these "instant reaction" thoughts may be internally generated and natural, but it seems possible that many might be remotely sent there by evil violent people on the other side, as strange as that sounds, that are trying to defeat their enemy which includes you and your friends. Probably a lot of the "voice in our head" and the feeling in our mind is sent there by violent criminals and is meant to mislead us and destroy our friendships. Sadly, there isn't really a lot we can do, other than to try to remember to ignore and counter as much of it as possible. So I find myself constantly reminding myself in thought-audio "ok maybe this person is not great, but violent people are much worse".

If you find yourself focusing criticism on allies (people with similar views as you have), for example for some trivial, non-violent difference, and not on clear enemies (those with very different views than you have: for example people who support the 9/11, Kennedy, Lennon, King jr. killings and lies), then probably it's the result of remote neuron writing. Many times there is some song in my mind designed to amplify some trivial difference among the anti-violent, and I try to quickly replace it with one of my self-made song melodies and lyrics, most often "9/11 was an inside job y'all", "tell the truth about the neuron writing", "stop violence, teach science", "journey to centauri", "sex ain't the big-azz problem, it's violence is the big-azz problem", "let the people get to see", "full democracy", "lock up the violent yeah, free the non-violent yeah", etc.

The one truth that seems clear about remote molestation and assault, is that it is a very good **opportunity** to fire back, in your mind, at some scummy person in the act of attacking you. Probably so few excluded people ever think about firing back, because they don't realize that some person is remotely writing on their neurons. In my experience, firing back in my mind helps. Not only does firing back erase the unnatural feeling the remote neuron writer gives you, but you have that wonderful feeling of firing back and really punishing some scum who is violating you and really deserves it. Any second, you have a bad thought, any second you feel an unnatural fear, any second you receive a suggestion to drive into traffic, or off the road, - any bad feeling or pain at all- it's an opportunity to fire back and really possibly punish some bad person who absolutely cares nothing about you, and is in the act of trying to kill or bother you, and deserves to be punished.

In some sense, firing back at people and machines remotely molesting us is like cleaning, because we can clean and everything looks nice for a few days, but we simply can't stop the unending accumulation of dust, and so we need to constantly be performing maintenance in the form of firing back when people remotely molest us. Finally, when you are too weak to fire back in your mind, the remote machines and people recognize that, and perhaps finish the job.

Remote particle molestation is probably a lot like feeding pigeons. A wealthy molester pays for a computer program to constantly send pop-up money D2B windows into the eyes of poor people with offers to "authorize" various remote molestations of their enemies, and there is a massive group of people who do nothing else but get money from "molestation money windows" (like the "shill money windows" many of us often hear the effects of) all day and night. It's not entirely clear, if those poor people doing the remote molestation are corrupted security people in the government, in the phone company, or in some other group. It seems likely that there are other "money" windows too; some might be for a pain, an assault, even a murder, and certainly many for pleasure-related services. It may be that these poor people receive some kind of fire back or punishment for their involvement in remote molestation- but the wealthy funders probably escape any serious punishment.

There is a good analogy between remote particle rape and a rape done at close distance, in that somebody is putting something in your body against your objections, and just like a rape done directly in person, as a victim, many times you naturally try to strike back during the rape to kill, hurt, mark, or somehow scar the person raping you. But with remote particle rape, for excluded people, striking back in self-defense can only be done by the victim through thought if even that.

Sometimes there will be a series of three or four molestations, like a bad reminder, you fire back, then the same reminder a second time, you fire back again, and then a third reminder- like a machine is doing the writing. It may be that many people reject the first suggestion, but eventually succumb to a second or third writing. Some of the suggestions are very subtle, for example, suggestions that simply trigger ideas, like suicide, or doing violence, without any actual images or sounds appearing. It is gruesome to say, but there is an interesting aspect that works to the advantage of the remote molesters, and that is that most people do not want to talk openly about thoughts of suicide or violence (or sexuality) that are remotely written onto their mind, because people might think that they are thinking about committing suicide or doing violence and panic, and so this significantly helps the remote neuron molesters to stay free and unseen. Just to add on this point, that I will never kill myself or do first strike violence against other people, I love life and have a very strong mind that is very familiar with remote neuron writing, so if ever there is some news story that I ended my own life or did first strike violence you can be sure that it was remote murder or assault. You can see how people need to accept a new paradigm of "the thoughts in your head, are probably, most of the time, written there by other people"- and then currently, by people who are some of the worst violent criminals ever to live on the Earth.

One positive result of the constant remote molestation is that, while completely unnecessary and absolute idiocy, it does make us build a stronger mind, and get better at firing back at molestations and bad suggestions with our mind.

Basic themes of remote neuron writing

There are some basic constant themes of remote neuron writing in the time we live in. One theme is to make the excluded people extinct. There is basically a "denial-of-service" (DOS) genocide of those who are denied the service of direct-to-brain windows and seeing, hearing, and sending thought images and sounds to and from brains. Not being able to communicate through thought with potential mates is probably the most effective part of the DOS genocide. Other methods are very subtle, for example to make the excluded fat by sending many food suggestions and therefore undesirable to mate with, and to make a person smoke tobacco lowering their chances of mating because less people will kiss them – in addition to the possibility of lung cancer removing the excluded. Another method is to make a person get hooked on drinking alcohol. A drunk excluded can be much more easily tricked to do something they will regret; some of the stupidest things I've done were neuron suggestions I bit on while drunk. Many of the suggestions are to stop an excluded from having the right words or confidence to get a date, or by making the voice of an unseen mate in the excluded mind make the excluded have a false sense of allegiance to the pretend mate, therefore rejecting potential mates that actually exist and are physically near the excluded.

Suicidal and violent suggestive neuron writings, are also another method to exterminate excluded, for example writing suggestions for them to jump off a steep edge, drive into traffic, drop their baby, throw their baby (you might think I am making this up, or that this is a ludicrous claim- but I don't even have a baby and I receive similar neuron writings all the time when I visualize holding my baby on my thought-screen– a woman who lives in the same county as me recently threw her baby off a parking structure – which certainly was remotely neuron written), or to use a knife or gun to do violence. Once you realize that these terrible suggestions are being sent by violent criminals, you realize just how sick and evil the neuron owners and many wealthy neuron consumers are. But those who are unaware, which is the majority of excluded people, have no idea that such evil people are writing to their neurons, and shockingly may interpret the neuron writing as being the wishes of a God, for example commanding them to do terrible acts of violence which they do because they don't want to disobey or be disrespectful to what they think is the voice or signs sent to them from a God. And the excuses given by these neuron criminals? "We done it for the right!" "It's right versus left!" "We done it for God, for Christianity, for Capitalism, for our side". There is always some label placed on the victim to try to justify the abuse: for example that the victims are perverts (this from multiple career remote writing voyeurs and molesters with victims of all ages), are insane (coming from people who reject the theory of evolution, and believe in the extremely unlikely claims of religions), are not 100% white, are gay, are not religious, are liberals, are democrats, are commies, or simply are excluded, they will never see us, that's just the way it is, it's free speech, and so on. Like the many murders, the Kennedy killings, 9/11 and 7/7: all bizarre neuron raping, assaulting, murdering, and molesting is forgiven, excused and forgotten by the D2B owners and a majority of D2B consumers.

Stopping the free-flow of information for excluded while simultaneously removing any clog on the free-flow of information for D2B owners and consumers is a common theme. For example, "rat" is a commonly written thought-audio (and paid-for shill too). One of the best thought-audio replies is to think "do unto others" because obviously they freely share every image of you. See my "Rebuttals" section for more classic replies from the victims. They freely "rat" with all your information- sharing every single image of your life and whereabouts with any and every direct to brain consumer- people with violence in their past, people who routinely remotely molest, uneducated people, religious fanatics, etc.- while many doctors, engineers, teachers, and self-educated people are excluded. The idea of a "rat" is absurd given the free-flow of info for the D2B consumers, but like the 19-hijackers story, true or false makes no difference whatsoever, only party line matters. In addition, much of this "ratting", for example telling people that all matter is probably made of light particles, that globular clusters are our future if successful, about walking robots with artificial muscles, and direct-to-brain windows is like "ratting" about some basic education- like teaching people that dangerous secret of "addition and subtraction", or "how to bicycle", or that secret "evolution theory" (imagine how people would leave traditional religions in droves if that precious secret truth were let out). Some of us may laugh at the thought of keeping the microscope a national secret, but yet, where are the low cost, hand-held electron microscopes by now?

Much of remote neuron writing is used to reinforce the claim that a person is "insane" or "weird". It doesn't take a lot of remote neuron writing to make a person do unusual things that might scare or make D2B consumers watching nervous about helping, befriending, hiring, or including the excluded. The thought-audio "psycho!", "cuckoo!", "loco!", etc. are often purchased and written to the neurons of victims. Much of the neuron writings are designed to try and spin the audience- to make those watching you entertain that question of "is a person a ...(psycho, pervert, etc.)?" – instead of more important thought-audio issues – like "murderers!", and "accessories to murder!', and "Turn off the D2B for the remote particle molesters and turn it on for all their victims!".

Echoing remotely written thought-sounds and mimicking thought-images

One clear aspect of remote neuron reading and writing is that many people who have never heard of such a thing, have a tough time remembering that other people can already see and hear their thoughts, and so they repeat "out loud" sounds in their mind, and "act out" or mimic images on their thought-screen. It's really unfair, because the D2B excluded don't get a constant reminder by seeing D2B windows in front of their and other people's eyes. Even I, who have had suspicions that people could hear and see me in my home, and possibly hear my thoughts, still echo thought-audio or mimic thought-images sometimes. It may even be that neurons can be activated or inhibited to stop people from remembering important things, like not needing to repeat written thought audio out loud. Sometimes a person may want to verify that they can correctly verbalize some vocal impression, or they may want to expand on some thought-image with muscular movement. But mostly, wealthy people who fund and control neuron writing have taken a really poor view of how to use it, and so, many sounds and images sent to a person's brain, are not in the best interest of the person to repeat out loud or act out. For example, many times a sound sent to the brain may be a song that may have a message or association that is not in the best interest of the person receiving the song directly to their brain (the audio might have a racist connotation, or imply that the singer is homosexual, etc.). Once a person remembers that sounds and images are being sent to their brain, which are mostly like "prank" phone calls, and not to repeat them, but simply to ignore them, then the thought-screen and thought-sounds become more like a detached screen and speaker, that you see and hear, but that are not your own thought images and sounds. Your thought screen and thought-audio are more like advertising "billboards" where mostly very crude wealthy people place "ads", mostly designed to make the excluded person echo some sounds that will make them look bad, or images and sounds meant to mislead them, or to slant the interpretation of events in a negative way. The wealthy molesters pay the phone company who then write these unwanted ads onto the poor excluded person's thought-screen and thought-audio without paying the billboard owner a dime.

External or internal neuron writing?

One classic annoyance is how a sound is sent to a person's thought-audio and it's not clear for any person listening to the victim's thoughts, if the victim thought it, or if it was externally sent there. Most of the time, in this terrible era we live in, the thought-audio is from an external source, and then is terrible audio- like racist words, anti-sexual put-down words, etc. It's like the rudest and crudest people are watching you and your friends, and sending the worst insults that come to their minds onto the thought-audio of you and your friends- like "you're fat", "you're ugly", "you're gay", "you're ..." on and on...with thoughts most average people would never think of. And the planetary excuse is probably always the same- why are the neuron writers so evil and annoying? Why, it's to fight against the enemy, of course! It's classical to send terrible audio to excluded people's thought-audio, and then those listening many times mistakenly believe that the poor excluded actually has that terrible view or belief. Even the excluded themselves may accept externally written thought-audio as if it was their own thought, and adopt the terrible belief or pickup and continue the conversation externally initiated in their thought-audio. The natural reaction is always to spend a few seconds thinking something in thought audio that will reverse or negate the message of the rude audio.

Using remote neuron writing to get excluded people to do violence

Trying to make excluded people do violence is one of the worst uses of remote neuron writing, and also very common. A suggestion might be as simple as activating a memory of a gun or knife the excluded owns, or saw once. The excluded then may entertain buying or using the weapon somehow for a few seconds. Many times there are suggestions to make a person hurt themselves, for example for a person to drive into oncoming traffic. The news is littered with stories of deaths that were most likely the direct result of remote neuron writing. Just in my mind now are a woman who walked her kids into traffic around Boston, another woman who drove her kids into a lake, all of whom died terrible deaths- and certainly remote neuron writing on excluded was what happened there – and certainly remote neuron writing was not used to stop them. It seems reasonable to believe that D2B consumers pay the AT&T D2B department to see these shocking videos of violence, as seen through the excluded eyes and as captured by microcameras floating around the scene, which I understand may sound unbelievable to many D2B excluded people. These videos are sent directly to the D2B consumer's brain appearing right in front of their eyes. So there is a simple financial reality: AT&T knows that they get a lot of money for violent videos, and so of course, they create violence through remote neuron writing then shop the videos to D2B consumers. The telecoms probably even get money from both ends- wealthy D2B consumers pay them to write violent suggestions on their enemies, and then they sell the videos for money to D2B consumers. Neither the D2B funders or owners are ever punished for any of the violence because they are too powerful. They must claim that what they are doing is protected as freedom of speech – that they are not actually pulling the trigger, but simply suggesting that the excluded person might want to pull the trigger, walk their kids into traffic, or off the road into a lake, etc. Like the bizarre story of the Old Testament of how God orders Abraham to kill his son (what wisdom, and decency there is in religions, huh?!), many excluded people think that the violent images sent to their thought-screen by evil people using remote neuron writing are a message from God, and don't want to disobey God, no matter how terrible God's request is. If only they knew about remote neuron writing, they probably would not act out and reproduce the remotely neuron written-on images of violence. Certainly much of the remote neuron writing is to try to get excluded people to do violence to themselves and to other people.

Even many of the other species are routinely made to do violence, presumably for entertainment and to create videos that people pay a lot of money to see. Classic examples are making dogs bite, horses buck (like for Christopher Reeve), bears attack, sharks bite, bees and stingrays sting (like for Steve Irwin), to get snakes, alligators and crocodiles to bite, whales to drag trainers underwater (like Dawn Brancheau), elephants to stampede, to get lions and tigers to attack (like the Siegfried and Roy attack), and probably many other remote neuron writings that cause animals to attack the many poor D2B excluded and consumers, who are viewed like dirt by many D2B owners and consumers. Often insects are remotely neuron controlled to collide into humans, as hard to believe as that may be.

It seems very likely that many vehicle crashes are the result of the criminal use of particle device technology, as unusual as that may sound to an excluded person. In particular flying vehicles are easy targets. It is probably very easy to remotely disable a vital part of any airplane or helicopter using the many microscopic particle devices that not only do RNRAW but that can do remote transistor reading and writing (RTRAW) too. Planes and helicopters should have safety devices like parachutes, airbags and emergency thrusters during this era where violent, lawless, unseen criminals dominate remote particle device technology. Given the truth about 9/11/2001 and 7/7/2005, how can we be sure that plane crashes like the famous Lockerbie crash aren't used to trick the excluded public into thinking that Islamic terrorists could do such a thing, to focus their anger against Arab people, and to help justify ridiculously enormous and wasteful military spending?

Much of the violence done is actually laughed at by many of the crude D2B owners and consumers. This is clearly seen even by excluded in everyday life on YouTube videos, and in popular cartoons like "the Simpsons" and "South Park" where chopping, shooting, and blowing people up is viewed as comedy- but even nudity is strictly forbidden. Many people in the bizarre violence-loving anti-pleasure D2B crowd laugh when their enemy suddenly falls and hurts themselves as a result of the invisible remote neuron writing. Part of the laugh response is probably relief that your enemy is weakened, but it may even be that an involuntary laugh response is written onto witnesses after some act of remote violence occurs. One vivid example of remote violence is how my Mom (an atheist who did modeling when she was younger), in her 60s, was almost murdered when she was remotely made to fall down the stairs in her home. An insider hinted to me that some D2B consumers laughed at seeing this terrible fall. One of the bones around her knee protruded through the skin as a result, and she lay there losing a lot of blood and too dazed to do anything. Fortunately a D2B friend of hers, probably broke with the tradition of non-interference, and went to her house, took her to a hospital, and my Mom slowly recovered over many weeks. Probably the "Christian" remote neuron writer(s) were not even punished for this vicious life-threatening assault, and to this day I have never even seen who it was that neuron wrote on her to make her fall, or who galvanized her to death a few years after that.

D2B Consumers sometimes make gestures and answer excluded people's thought-questions

Of course, you can think questions in thought-audio to the D2B consumers, which they definitely hear. It is somewhat unpleasant for them to have to respond with some kind of muscle movement, a nod, a vocal "yeah", a movement of the eyes, or fingers, etc. – since excluded people are prevented from simply hearing any thought-audio answer. Bad D2B consumers take the many D2B money offers to mislead excluded people, but decent D2B consumers may reject the thousands of dollars and honestly answer questions asked in thought-audio by excluded people. It may be that D2B consumers get some party-line answers from a D2B service provider with more authority, so that they themselves are not labeled a "rat" by responding to the thought-audio of an excluded person.

One classic question to think is: "Do you get D2B?" – many people do, and most often they turn their head to indicate "no", which probably means a variety of things: 1) yes, they get D2B, 2) no they don't think it should be public, 3) they want nothing to do with you, 4) they at least have the courtesy to indicate to you that they receive D2B, or possibly simply that 5) they are too afraid of "Orwell", the neuron owners, and their capo-like thug employees, to do anything other than indicate with a "no". Since head movement can be triggered in excluded by remote neuron writing – it leaves an excluded person without any real certainty about how many people get D2B and how many don't.

Remote neuron writing on D2B excluded is very fast

Much of the remote neuron writing that excluded people get are images and sounds that last only a millisecond, but then resonate for many seconds, because most excluded people don't realize that you need to fire back and try hard to erase them out of your mind. As an excluded you don't get long duration, easy to see, D2B windows, all you get are unpleasant flashed images and brief sounds, and then mostly on your thought-screen, and rarely on your eye screen. A good example is that you heard a voice in your thoughts, usually a rude one word put-down like "perv", etc. that didn't come from anybody around you. Another example is when you thought you saw a snake, or a spider, or somebody you recognize but then when you looked again, it isn't the same image. Much of the writing on D2B excluded occurs when they are going into or out of sleep. In addition, a lot of remote motor-neuron writing (remote muscle contracting) is done to those who are excluded as I have already said.

Many times the D2B payers send an image, sound, or activate a memory for a millisecond and then must watch the excluded thought-screen and listen to the thought-audio to see what the excluded person thinks, while the excluded is ruminating on the millisecond image or sound, many times in a negative way. It's very tough at first to quickly dismiss the remote neuron millisecond writings or to quickly draw thought-images and thought-audio in opposition or to change the subject. Here's a typical scenario: Wealthy mass murderers want to rip apart their enemy, progressive, honest, educated and anti-violence people, so they send an image of a popular progressive onto the brain of an excluded progressive, and then pump negativity onto the excluded neurons- it's designed to cause chaos among the honest and fracture stop violence alliances- not just among progressives – but among any honest and stop violence people.

Many neuron writings are ads, in particular for food, smoking, alcohol, drugs, and other products, to try to make people make decisions – many times decisions that are not in their best interest. Many excluded might be recognized because they are: obese, smoke, drink alcohol, have theft, sexual, or violent offenses, etc. because they don't realize that D2B images, sounds and neuron activations are just suggestions and not to bite on them. People probably pay the neuron writers millions of dollars to write to people's thought-screen billboard, and radio-station-like thought-audio, which many D2B consumers may pay to watch. But the owner of the neuron does not receive one penny from that neuron writing rape. It is a simple idea that I think many people can agree on: that there must be consent, and that people should pay the owner of the brain to write advertisements directly on to their brain. The current system views excluded people as walking machines that can just be written on without permission, and without paying a penny to rent space on their thought-screen billboard from the millions of dollars the telecoms are paid to write ads and other unwanted messages there. Most of the direct-to-brain writing, from my own perspective as a person who is being denied D2B windows, is unlike television and radio, in that direct-to-brain writing is a service that we cannot turn off or on.

There may even be *advertising* on D2BW videos of excluded people in order for AT&T to get even more money out of the remote one-sided rape. How many D2B consumers find themselves thinking "I paid to see Ted, but all I see is Tide"? Adding advertising videos or images on the side, or before seeing excluded people in a D2B window, may make the remote gang rape feel more like television or YouTube, and less like the gang rape of the poor excluded people that it is.

Many of these remotely written millisecond images on the thought-screen of the excluded are of people that they don't want to think about, or trigger unpleasant memories. So life for many excluded becomes based on how quickly they can forget that latest neuron write. These sounds and images sent to people's thoughts are definitely a form of molestation, neuron rape, and prank calls – purposely sending a memory, image, or sound that annoys you. Some of the millisecond images and sounds an excluded may receive may actually represent a pleasant communication, although this is the extreme minority- perhaps about 1%, if even that. It's like the extremely limited D2B "service" that a D2B excluded routinely gets. One comical phrase is that the excluded person is just really "far out". Instead of an actual D2B window with a video, all the excluded gets is a millisecond image or memory triggered of some person. The excluded has no idea if that image was sent by the person in the image, or by somebody else. Instead of regular windows in front of all of our eyes, all we D2B excluded get is a millisecond image flashed in the back of our mind, or just a millisecond memory of something we saw in the past – the vast majority sent strictly just to annoy us, to waste our time, and to exploit the many people that might be watching the thought-screen of an excluded – with no thought of compensating the owner of the thought-screen-billboard. So in theory some (<1%) of these millisecond images could be actual thought phone calls from poor D2B consumers who are usually not allowed to call you. In addition, because AT&T and Verizon have a monopoly on D2B (at least in the USA), probably the price to make thought calls is much higher than it would be if D2B was democratic.

Why does the phone company, or the neuron owners, or whoever, send unwanted images and sounds to the excluded? What idiots pay the phone company to send these annoying thought images, sounds, and memories? Many times the annoying memory has to do with some victory for the bad people who send them, for example activating the memory of some unpleasant experience from your past – an argument with a loved one, a memory of something you did that was unethical, etc. Many remote molestations of excluded are probably meant to make an example out of an uppity excluded – to say to those watching: "look what we do to this excluded", you don't want to cross us or we will do that to you too. But also, many must be to show the audience watching the excluded person that the excluded really is a second-class citizen – not an equal – not heroic, to spin the opinions of those D2B consumers that watch videos of the victims, or perhaps just to make watching popular excluded people annoying to all involved, in particular to the poor excluded being neuron raped. It's all part of a massive war of evil against good. Many times the remote neuron annoying writing is meant to try and divide the honest, which form the other side or the "enemy" to the many violent criminals that have access and enough money to pay for remote neuron writing (like those who support the 9/11 controlled demolition). One example is sending you audio in the voice of your friend or loved one that puts you down or ridicules you in a shallow way. It's similar to the Popeye "Hold the Line" 1920s video. The strategy is to make two bunnies become distracted by fighting with each other so that the fox can come in and eat both. The evil of the D2B consumers and owners have this "divide and conquer" strategy down to a science, and it is very effective- in particular against the excluded public.

The vast majority of remote neuron writing that excluded people get are "identity theft" (again, just like the Popeye "Hold The Line" cartoon), and they are meant to break up or stop friendships, relationships, marriages, and families among the "enemy" of the wealthy who own and control much of remote neuron reading and writing. Some examples include sending sounds of the voice of your wife or friend saying "go away!", "fat!", "shut-up!", "whore!", "gay!", or "I hate you!", etc. – but the actual person never said or thought that. Other examples are sending sounds of a spouse or mate having sex with different people – including enemies, friends, coworkers, etc. - that never happened. Identity theft plays a large role in 9/11, 7/7, and many other violent acts: a group of wealthy people use military people to do the actual violence, but then buy news stories that blame excluded (in this case Arab) "patsies" for the violent crimes that the wealthy D2B people and their employees actually committed.

Much of remote neuron writing is "slanted" to the violent and dishonest criminal view

Most, perhaps 95% of remote neuron writing that is not motor neuron unwanted muscle moving, is extremely slanted toward the violent and dishonest side, because that is who owns and controls much of remote neuron reading and writing. Just like there is a "best of" collection for some music group, what most people get written to their ears is like a "worst of" collection. All we ever hear are put-downs, and other misinformation from the lowest and most far removed from factual minds. To a large extent the voice and images in our head may be a direct representation of the views and hopes of the D2BW owners since, ultimately, even beyond those evil D2B consumers who buy annoying neuron writes, they, the owners, ultimately decide and control absolutely anything and everything that gets sent to any brain directly. The views are always far from your own views in being crude, rude, always putting down the receiver of the neuron write in order to make them feel weak, ashamed, flawed, etc., but also some times to "appease" the receiver into happiness with false promises and allusions of pleasant futures that are, in reality, very doubtful. A good thing to realize when you receive thought-audio, images, or feelings, is to picture that a really evil Nazi is the only person at the controls that can write to your brain- because the truth is probably not far away from that.

When they write rotten audio on excluded brains or as "news" stories it is "free speech" and "free market", but when you want to buy ads and widely publish images that show and tell all about remote neuron reading and writing and all their violent crimes, it's "ratting", "crazy", and a violation of "national security" and "privacy".

Planetary Good versus Evil?

Is there a constant two-sided war occurring around many stars like ours? Clearly there is a spectrum, with stopping violence and telling the truth on one side, and doing first strike violence and lying on the other side, but perhaps this is too over-simplified. To survive, some matter is always going to eat or separate some other matter. Just like there may be a planetary alliance to stop violence and teach science, so there may be an opposite alliance that seeks to start violence and stop the teaching of science. Just as there are many people who are in favor of the public being shown and told the truth, so there must be an opposite group of people who are in favor of hiding the truth and lying to the public. Our species defines "good" as stopping violence, teaching science and telling the truth, and "evil" as starting violence, and lying. Using this definition, good and evil, like ying and yang, pleasure and pain, or sweet and sour, may forever be in a classic struggle that never ends.

The dominance of religions with their wildly inaccurate beliefs, and many brutal murders of history, over the minority of science is a clear theme of most if not all of recorded history. The massive and long lasting lies like the secrecy of remote neuron writing, is a clear indication that currently on our tiny planet, evil is in the dominant position. Why have the evil (those who do first strike violence and lie) been so "large and in charge" for so long? One mystery is that since they are almost all a "white-men-only" ultra-elitist club, how can they possibly maintain a majority and stay so firmly at the top for so long? Wouldn't those who are for gender and race equality in law and for integration of the many races and people form a much larger and wealthier group, because they are not so exclusive (they include women and non-white people)? For whatever reason, this apparently is simply not true yet.

Because of their great wealth, the views of the all-powerful evil at the top tend to be over-stated and dominate everybody's eyes and ears. Wealthy people who are good people fear confronting the violent, so the evil of the wealthy people tend to dominate. Their focus is always on trivial, petty, and non-violent activities, like sex, drugs, mental instability, etc., while the views of our group, on the opposite side, tend to be extremely understated, and focus on only the most serious of the problems. For example, we are simply trying to even tell the public the truth about the centuries of remote neuron reading and writing, and about the many wrongly "solved" crimes like the controlled demolition murders of 9/11, which are not non-violent petty crimes where everybody lived and nobody suffered any pain, but instead, shocking, painful, and terrible pre-planned successful murders of thousands of innocent people.

On a positive note, we are all in a competition to see who can expose, weaken, and punish evil the most. Those who everybody recognizes as the most effective at exposing all the lies and violence of the evil people will probably be the most rewarded and respected if ever this current group of evil collapses.

"The Shill", "Money windows"

As a D2B excluded, it is very difficult to understand that much of what happens around you, what people say, and who is walking around you, is strictly the product of a payment of money. One way that the Internet is different from the D2B net is that in the D2B there are, apparently, many "pop-up" windows that are like "Vegas-style" offers for money. Mostly, the pop-up windows pay the D2B consumer to say something abusive to an excluded person (like "perv", or "gay", etc.). Some D2B consumers hint that this money is not like $100 or $200, but is more like $800, $1000 (a "k"), and more. I can imagine that it is a lot of money, because those who pay have a lot of money- to them, in particular, working collectively, a thousand dollars is like nothing- and because average people probably demand a lot of money to take the risk and potential abuse that might result from saying something abusive to an excluded person- after all you basically are probably going to destroy any friendship and kindness that might have existed before the shill. But the money box pays people for more than just verbal insults to be sneakily hurled at passing or stationary excluded- there are all kinds of offers- D2B is a rapid moving free market for those who are in it – similar to EBay with many people bidding up and the offer price constantly going up or down. There are not just money boxes for saying things, but also offers for sex (hands, sucks, coatings, coverings, holes, etc.), for dates, to physically go somewhere for money, muscle moves, no doubt assaults and even perhaps murders … I can only imagine.

But just in terms of the money box for "shilling", there is a multi-billion dollar industry of misleading the excluded. Your D2B "friend" is almost always more accurately called your D2B "fiend" – but there certainly are those who refuse to say things for money. Shilling covers the entire spectrum: Those who refuse to take money to say something, those who say somewhat helpful things for money, those who say somewhat neutral things for money, those who say somewhat slanted/mildly abusive things for money, and those who say extremely slanted and/or abusive things for money. Probably a common thought for excluded people is "why are so many people so shockingly rude to me all the time?". The answer is simply because they are paid at least $1,000, and perhaps even $2-3k for each rude remark. $1k and more can go a long way for a poor person- who needs to pay rent, tuition, loans, etc. With that kind of money, the anomaly are those people who turn down the money windows to shill – they must be going for a higher and more long-term money that comes with good credit and non-abusive behavior.

But even if paid, why would people pay somebody to be so rude? For many reasons: 1) they want to protect their position of wealth by putting down popular people who oppose them, 2) to stop their violent crimes from being the focus of attention, 3) to isolate excluded people by making friendship with a D2B consumer less likely, and to solidify the segregation – the myth that D2B consumers are superior, and that dumping on and viewing excluded people as inferior is ok, 4) for sexual reasons- they may be aroused by a person abusing an excluded, 5) It's a way for them to express and promote their viewpoint, and party-line to the excluded and those watching, 6) To confirm myths and lies that they have created or maintain: that 19 hijackers did 9/11, that Oswald killed JFK, that there is at least 1 God, that their excluded enemy is a pervert, gay, and/or has a mental problem. Ultimately shilling is because many D2B owners and wealthy people that fund shilling are just low-brow people. But while the people who fund shilling may be low-brow, the technology behind it is sophisticated.

Sophistication of the Shill: The D2B computers are smart and try to find words and phrases that a D2B consumer will feel comfortable taking money to say to an excluded. Many times the word will appeal partially to something they agree with, or that has a double meaning that satisfies two opposite sides. Mostly, though the shill always favors the D2B owners, and most of the time expresses a crude and/or distorted, dishonest, and deceptive view, which is the nature of buying a shill – the reason the D2B consumer needs to be paid is because they probably would not normally say the paid-for word or phrase. Shilling can be more than just a single word, for example, a person might be paid to say a complete sentence. Another tricky example is that a supervisor may be paid to harshly criticize an excluded employee under them, for example saying "what were you thinking?!" or "why didn't you do that?!", because it may lower the confidence of the excluded person. Sometimes people may be made to shill without being paid, for example if their muscles are moved remotely involuntarily, or when a shill word or sentence is remotely made to feel desirable to say. And the paid-for unwanted sounds to our ear don't stop when an excluded person walks away from a D2B consumer, because the shilling continues with paid-for audio sent directly to your ear neurons. Like a prank phone call, the audio usually takes a nasty and dishonest view, deifying the neuron, demonizing the excluded, deifying remote murderers, assaulters, molesters and liars, demonizing those trying to expose and punish them. Like a prank phone call, only the phone company knows who and where the sender is, and they don't tell the unconsensual receiver. An excluded can certainly try to think "no call list" or "out" in their mind, but it's probably useless; because for the neuron, money and their personal views are the bottom line, not majority views, fairness, decency, honesty, law, stopping violence, teaching science, etc.

Another aspect of the shill is the personality shaping that it does. Just like Pavlov's dogs, human behavior is shaped, not with food as much, but with D2BW money. D2B consumers are trained using fast payments of money to fulfill the desires of D2B owners and consumers. Not only are there money windows with offers to say rude words and phrases, but there must be a lot of "teleprompting" and "script reading" – phrases and words are offered by the neuron, just like the Internet, to help guide D2B consumers in their choices of what to say, what to see, how to win friends, how to deal with excluded people, how to lie about D2B, what to say to pretend that they know nothing about D2B or anything else, what to do, what to think, what opinion they should have, who to vote for, what products to buy, what songs to listen to, etc. – and then all with a little money "treat" if they obey probably. I've seen people young and old grab that $1000 or whatever in their eyes and just shout out words at me like "go!" and "perv!", in an airport, or walking around a shopping plaza- the Earth has become just like a virtual rude Nazi Vegas.

There is something so despicable and low-brow about tricking and abusing a deaf and blind person, it's like taking candy from a baby, and this is exactly what so many D2B owners and consumers engage in every day. They prey on the weak and those with the handicap (which they unfairly and needlessly create) of being excluded from D2B. D2B excluded people can't see and hear thoughts like they can, and just like deaf and blind people, excluded are easy targets who don't offer much resistance – they are the weak lambs of the herd. Remote neuron writing onto excluded people is very much like a rape because the excluded person does not ask for the writing, does not consent to the writing, and has never even been told that remote neuron writing exists and is being used on them.

Possibly one added aspect of the constant popup money window is to stop the D2B consumer mind from wandering or having time to direct their own thoughts.

One of the best ways of looking at those who routinely shill, is to see them as what they are: kind of like a robot that people put money into to echo statements out loud- and just to ignore them and what they say as quickly as possible; to see them basically as meaningless money machines. Some D2B consumers can give good advice, but the vast majority of money windows promote a slanted view. Obviously the best advice to D2B excluded people is not to look deeply into the words D2B consumers who shill a lot say. The more the excluded trusts the D2B consumer and views them as a "friend", the higher the money that is offered for the D2B consumer to shill and mislead the excluded person.

People are just like a share of a company stock; their value goes up and down. When they take money to shill, their value goes down because nobody wants to hear shilling around them. When I hear a person shill I many times think "hold out for the big money...that comes with honor, honesty, decency and integrity". Sometimes I think "you might as well put a negative sign in front of the money offered to shill, because that is what you are really getting".

The best thing about a clearly stated shill is that, you know, for sure, without any doubt whatsoever, that the person shilling definitely gets D2B windows. There is no question about it. They are not one of the millions of excluded victims that need to be told about and shown D2BW. That's a valuable piece of information for the D2B excluded. In fact shilling may be the best evidence for an excluded person that secret remote neuron reading and writing (and D2BW) even exists. Thank goodness for people's constant desire for money, because it provides some of the best physical evidence for RNRAW and D2BW. Excluded may even accumulate numerous videos of people shilling to present as evidence of remote neuron reading and writing in a court trial (with an excluded judge and jury perhaps), or to help educate and inform other excluded people about RNRAW and shilling. Another consolation is that they only lose money when they shill. They gain a short term thousand, but their credit goes down, and in the long term nobody wants to hire or date them because it's unpleasant to be with somebody who takes money to say abusive words. Sadly, to a certain extent, those poor dumb D2B consumers who say unpleasant things for the idiot wealthy people who pay them, are being naturally selected for survival in being rewarded for abusing and misleading excluded people for money. Beyond that, on a positive note, you can think that any time a person shills at you, that you might be getting richer, because that money may ultimately be transferred to the victim of the paid-for abuse. It's a tough life for many poor people and it's good to put everything into perspective, shilling is nonviolent and free speech; violence is the big evil, shilling is just annoying. Hopefully people will vote and all the money gained from shilling will go to each particular victim.

In my experience, one characteristic of many D2B consumers who shill is that they can "dish it out, but they can't take it"- many times if ever I return a shill verbally, it usually has a dramatic effect on the person who shilled- they can get very angry or upset. The key is understanding the dominant view of people who are being denied D2B windows by those who do receive D2B windows; those denied D2BW are viewed similarly to those accused of "witchcraft" and "blasphemy", or like native African people in South Africa under apartheid, Jewish people in Germany under the Nazis, or the "untouchables" of India. Each nation apparently has similar analogous groups; people who are viewed as second class for mistaken, petty, and/or sinister reasons. So a D2B excluded "talking back" as if they were an equal to a D2B consumer can be unusual and upsetting. The view is just like there is an "uppity" or disrespectful slave, but the truth is that the shill deserves to receive the same abuse that they deliver to their excluded victims- most of whom are excluded not because they are somehow flawed or inferior, but probably much of the time, because of their unwillingness to lie about D2B, about scientific truths, or for their criticisms of popular mistaken religious beliefs. The D2B included can be shockingly rude; they harass and bully, many times receiving thousands of dollars in the process, but if a D2B excluded ever fires back, or "returns a shill", all the blame is put on the D2B excluded. It's a lot like a racist saying "did that slave talk back to you?!".

As a D2B excluded victim of shilling, I encourage you to be better than the people that shill, to rise above, to not become angry or violent. Just like any victim of a corrupted "class" societal structure where fine and wonderful people are treated like dirt, don't sink to their level, instead, focus on your goals and dreams. It very well may be that every single penny (ESP!) gained by those who shill may someday be paid to the victim of the shill. And when I say that you, as a D2B excluded, have to be better than the D2B owners and consumers, I mean that you have to be 1000 times better, because just like in any place where there is an unfair segregation, the second class must always be 1000 times more friendly, talented and productive to get a job, a room, a mate, etc., even though they get 1000 times more abuse, and 1000 times less information, physical pleasure, reproductive rights, money, job, and housing opportunities. Plenty of wonderful, beautiful, brilliant, skilled and talented people have been burned at the stake, gassed in death camps, galvanized, etc., because they were labeled second class citizens by the dominant uneducated wealthy D2B big-lie murderer elites.

Sophistication of the remote neuron reading and writing technology

If remote neuron reading and writing is 700 years old, then clearly those who own and control it have had a long time to develop it, and it shows. There are many aspects that, as an excluded, I certainly am unaware of. But, one thing seems clear, and that is that the owners of RNRAW can render many objects in three-dimensions and in real-time. It seems likely that all 7 billion people are modeled in 3D and their motions predicted. The computers probably predict the motions of humans and other 3D objects generally over a long term- with regular activities like when we will go to our job, etc. – but also into the near future, perhaps even hours into the future – for example knowing when we will reach any particular stop light, etc.

Defend yourself against the remote neuron writing

For the love of life, defend yourself against biting on bad suggestions. These remote neuron writings can take the form of barely visible images sent to your thought-screen, barely audible sounds sent to your thought-audio, and memories that are triggered in your mind – there are many subtle and sneaky remote neuron techniques. I know from experience that sounds played to the ear and thought-audio of a D2B excluded can be misinterpreted as "the voice of God" in your thought-audio. I made this mistake – and I'm not religious! Think of how common the phrase "heard voices in their head" is. Many times, this is what those who have murdered somebody say. Mark David Chapman said how he heard "do it! do it!" in his thought-audio before he killed John Lennon. Recently a person jailed for the murder of homeless people, slammed his head against part of his cell trying to make the voices in his head stop. You can see how a person who sends suggestions to a D2B excluded person might claim "I never pulled the trigger" and plead "free speech". Getting an excluded person who owns a gun or other weapon to do violence is a classic remote neuron writing. Much of this shocking and disgusting violence and abuse of excluded pawns would end if only the public was even just told and shown the details of remote neuron writing.

One possibly helpful idea is to "fire back" in your mind at the remote molesters. I find myself at the "fire back" laser in my mind every 5 seconds. Imagine those that don't fire back, but readily accept the terrible remote neuron written sounds, images and other suggestions- they are easier targets and just like typical predators, the neuron rape writers pray on the weakest of the herd. The feeling of "stage fright" can be remotely sent. Even simply "fear" remotely stimulated in your mind can be remotely sent – for example when you turn out the lights before going to bed. I find that firing back in my mind helps tremendously. Let's hope and vote that the remote particle molesters must lose their neuron writing permission. In addition, they should lose their neuron reading permission – at least until neuron reading and writing is public and their victims can read and write from and to their neurons too. Certainly I think that their particle molestation should be done back to them. One idea is to think a response that is similar to the molestation being done to you- if you are being made to feel like you are getting an enema, think about them receiving the same sensation. I think the key is not to become angry and in particular not to become violent. It may be that the images and sounds victims of remote particle molestation think in response have almost no significance at all and that much of remote molestation may be to try to make a person "blow a fuse", become violent, and attack people who, unlike the remote molesters, *can* be seen and *are* within physical reach of the victim.

You might think that a wealthy people that own and control remote neuron reading and writing would be "good parents" to the excluded "children" of Earth, that those with the immense power and advantage of remote neuron reading and writing, would at least, since they refuse to openly describe and share the technology, show responsibility, and be like a helpful "seeing-eye dog" to those excluded they have made "blind". But experience has shown that the exact opposite is closer to the truth. If they are a parent, they are a parent who constantly tries to find new ways to kill their kids. If they are a seeing-eye dog, they are a seeing-eye dog with a death-wish for its blind owner. The focus of most remote neuron writing is on exterminating, assaulting, molesting, and misleading most of humanity.

Figure 5.2. What looking at people with Direct-to-Brain Windows enabled might look like. Notice that the dog (that receives D2B windows) see's part of it's nose on it's eye-screen. Notice that the only people with friends or babies are D2B consumers.

Included and Excluded

There is clearly a segregation of those who get direct-to-brain windows (many times called the "included") and those who have never even heard of direct-to-brain windows or remote neuron reading and writing (the "excluded"). There will always be inequality among humans; those who have more money, those who are younger, better looking, smarter, etc. So in some sense, for a D2B consumer walking around with D2B windows seeing many other people without D2B it may be like the way those with homes look at homeless people – with pity, but an acceptance that some people are always going to have less and have a tougher life than we do just like others have more, and have an easier life than we do. But the D2B segregation is more like the segregation directed at Jewish people in Nazi Germany, or African American people in the United States under slavery; it goes far beyond just a difference in income or appearance, and is, I think, in large part, an "ideological genocide", based on a person's religious, and political beliefs, how honest a person is, and probably to a large extent their gender and race. Then to not openly acknowledge this massive secret system, even to the extent of not allowing a D2B consumer to admit that they receive D2B, is clearly evil and unnatural. One of the worst aspects of the D2B segregation is that, unlike the openly debated segregation based on gender or race which is often documented publicly in laws, like the famous Nazi laws forbidding interracial marriage, it is almost impossible for the D2B excluded to "see on paper", or to even know that they are being extremely discriminated against, and being remotely abused.

Denying millions of people knowledge and regular use of D2B creates an imposed disability and handicap in millions of otherwise healthy people

The owners and controllers of the vast nanometer sized particle (neuron) technology have created a handicap in denying many people the regular service of direct-to-brain windows. Because so many millions of people do receive regular D2BW, it creates a disability for those who are denied-because modern people are communicating through thought-audio and sending thought-images back and forth, but the excluded must be told out loud and have a physical image shown to them to understand what has happened. D2B excluded walk around and grope in darkness while the D2B consumers just watch and worry what the poor excluded people might do. That excluded people are not even told that many people see and hear their thought images and sounds, and that even telling them this out loud is forbidden by the neuron owners adds an extra shockingly callous and inhuman dimension to the imposed disability. I can understand denying people with multiple counts of violence the right to receive D2B, or even removing the D2B write permission for those with multiple remote particle molestations, but most of the people being excluded from using the "thought-phone" or "thought-net" haven't committed any violent crimes. The popular view among D2B owners has traditionally been one that favors an ignorant, uninformed, deliberately deceived, and uneducated public. But, I think the opposite view, informing and educating people, is the path to safety, survival, and physical and intellectual pleasure. Knowing that our fate is to try to build a globular cluster, serves as an incentive, helps people to see the big picture, and they then shape their lives with that in mind. Probably many mistaken scientific beliefs and barbaric religious traditions would come to an end if everybody learned the truth about RNRAW. Everybody knowing that thought can be seen and heard would vastly reduce the needless use of paper and ancient technology, in addition to the time saved in not having to repeat everything out loud again. The free flow of information always helps because the majority can see what the problems are and work to solve them, as opposed to groping in darkness with mistaken beliefs while those causes of damage continue to grow unstopped. Beyond that, communicating through thought-audio and thought-images has become so common that it should be made public, like the cellphone, and like a cellphone and the Internet, middle income people should be able to use D2BW for a low monthly cost, in addition to receiving a percentage of the money paid by the public to the telecom companies to see them and their thoughts. We're hurting ourselves by keeping remote neuron reading and writing a secret because analogous species of other stars are not delaying, as they continue to grab more stars.

D2B consumers have to pretend

One D2B consumer kindly said to me "We have to talk", which is a very important truth of this dark and medieval era- D2B consumers can just think to each other (sending thought-audio and images back and forth), but have to talk out loud to D2B excluded people. If everybody could regularly see and hear thought, obviously there would be no need to repeat thought-audio or describe images people see in their eyes or thoughts out loud. Making D2B consumers have to talk out loud is a big waste of time and very annoying. The current rulers of the neuron insist, as generations before have, on maintaining the myth that D2B consumers cannot see thought-screens and hear thought-sounds. For example, if there is a meeting at work (fig. 5.1), and all the participants are D2B consumers, they still MUST talk out loud – they cannot simply conduct the meeting in thought audio and image. Is that not absurd? You can imagine society on Earth if and when the segregation is public and D2B consumers hold meetings in silence, occasionally talking out loud for the benefit, or to explain something to a "deaf" excluded person. Perhaps the excluded people will wear headphones and the D2B consumers will type their thought-audio notes into a synthetic voice machine for excluded people forbidden from receiving D2BW to hear with headphones.

D2B consumers have to constantly "play dumb". They know everything about the excluded, but when the excluded asks "did you know this?" the D2B consumer must always pretend to be surprised and have no knowledge of events at all. The "playing dumb" goes to a ridiculous extent, comically parodied in the movie "JFK II"[123]- I can imagine D2B consumers saying "what's a neuron?", "what's a particle?", "what's a JFK?"... just endless ridiculous denial of obvious truths they all know in great detail.

One great annoyance is that D2B consumers cannot respond to the thought-audio of a D2B excluded who directs thought-audio at them. The D2B consumer cannot answer any thought-audio questions out loud, but some may nod or make other subtle gestures or use coded words to respond out loud.

---

[123] John Hankey, "JFK II" , 2003.
http://topdocumentaryfilms.com/jfk-ii-the-bush-connection/

Who gets D2B and who doesn't? Who is included and who is excluded? Who is "innie" and who is "outie"?

One of the classic questions of these centuries is: "Is so-and-so included?". For example "Is my parent included but I'm not?", and "Is my child included but I'm not?". It's unbelievable, but yes, in many families, one or more members may be excluded from even knowing about D2B, while others receive regular D2B windows. In my own immediate family both my Dad and brother are "included", and receive some regular D2B windows (although as excluded, you can never be 100% certain), but my mother was never included. Those family members who do get D2B are absolutely required to lie if ever asked if they see or hear thoughts. However, many family members do care about excluded family members and so are constantly "hinting". Most of this hinting takes the form of key words. Hinting is acceptable, but any kind of direct and open acknowledgement of getting D2B is absolutely forbidden and must be severely punished – probably by losing D2B service for long periods of time, and perhaps permanently. An included uncle of mine hinted to me about remote neuron reading when I was a teenager- only much later in my 30s did I realize the importance of what he was telling me about pilots controlling a plane using only their thoughts. For excluded people that know they are excluded (that "try to use the phone" as one D2B consumer put it nicely- they direct thought-audio at people), hanging out with D2B consumers, in particular D2B included family members, can be a tough situation, because the excluded who knows about D2B usually trusts their family members (unlike many of the people who shill around them) to tell them the truth, and tries to get some answers from them. D2B consumers see so much more than excluded people and so have a much better view of the big picture surrounding an excluded person. As a D2B excluded that knows about D2B, all I do is think thought-audio questions to my included relatives, and all they can do is just hint by nodding and hinting out loud by saying "**yeah**…I think this or that", or "You **know**, I …", and by using other creative word choices. Once, for example, I asked in thought-audio – "What did Mom think about a lot?" and the answer: "mao"- which implies lots of images of food (eating food is sometimes called "mowing food")- my Mom loved all things food, and plus there is a double-word score with the word "Mao" for "Mao Zedong", the famous leader of Communism in China; being atheist, as my Mom was, instantly earns you the label of "Communist" even though my Mom was outspokenly for Ayn Rand style "laissez-faire" free market capitalism- she had never even heard about *full democracy* where the public gets to vote on the laws they are subjected to. Be aware that, even your immediate family members will deny and lie about receiving D2BW to you if confronted directly, in particular about seeing and hearing your thoughts. Whatever makes them lie is apparently a very scary threat or something that would make any average person lie to even their own mother. In addition, there is the addiction and absolute fear of losing D2BW service. There is a clear analogy to being locked in a prison camp: pretty quickly, you realize that to survive you need to say whatever the camp guards tell you to say no matter how untrue.

Are there any guides or techniques a D2B excluded person can use to determine who is in and who is out? Believe it or not, there are. If a person "shills" you know they are "included" (they receive D2B), but you have to really hear a number of clearly spoken shills to be sure. This is one really good reason that an excluded person should not become angry or emotional when hearing a shill, but instead, recognize the great value of knowing that a person gets D2B. If the shills appear to be very quick involuntary "Freudian slips" – then the person is probably excluded too – because a shill takes preparation, where an involuntary motor neuron write is too fast to be pre-planned. Some D2B verbal statements for money are hard to detect, because the D2B consumers definitely want that money, but they definitely don't want to say something offensive or mean to the excluded target. The shill payers definitely want to direct the topic of the thoughts of excluded if they can, and so the shill-for-money words may often seem somewhat neutral. Terrible D2B consumers are easy to detect, because they always take the money to say clear and terrible insults, but decent D2B consumers will usually let you know that they see with some kind of words, in particular like "I see", or a gesture like holding a fist or finger to their mouth or eye. The absence of a shill is one of the best indicators of an excluded. Think of those around you who have never shilled-they are most likely excluded. The only other alternative is that they are D2B consumers that refuse to take money to say things to excluded people. Most D2BW consumers say something or make some clear gesture which reveals their D2BW reception pretty quickly. If any person deliberately covers their mouth with their hand, or makes some kind of deliberate and meaningful hand gesture just when you first see them, you know instantly that they are probably D2B consumers. The number one give away of a D2B consumer is the shill – it's driven by money and so many D2B consumers bite on them. In fact, the excluded may form a searchable Internet database or "Wiki" that identifies each human of Earth by name and photo, and provides video evidence of shills and gestures, or from thermal imaging cameras, that indicate if each person is a "D2B consumer", or a "presumed D2B excluded".

Some people may nod their head in response to your thought audio question. If they do then probably they are a D2B consumer, if they don't, then they are either too scared, don't care enough about you to respond, or are excluded.

A D2B consumer will generally not look at an excluded, while an excluded probably will look at another excluded if passing them. A D2B consumer doesn't have to look at anybody around them, because they get D2B windows with real-time video from nearby floating wireless nano-cameras, but D2B excluded have to look around to see the people around them. In addition, a D2B consumer usually wants nothing to do with an excluded person, because communication is impossible, because there is no chance to get anything (like money or a job) from an excluded, and because there is a fear of an excluded acting out remotely written bad suggestions. So a D2B consumer will not make eye contact with an excluded because 1) they already see them in their eyes, 2) a relationship would be impossible and annoying, 3) there is nothing an excluded person can offer them, and 4) they fear excluded people as dangerous and more likely to act on bad remotely neuron written suggestions.

Is one of my friends included? Mostly D2B consumers only befriend other D2B consumers, but they may have a "hi/bye" relationship with an excluded. It's very tough for excluded people to make friends. If they have a lot of friends, probably they are a D2B consumer. If they are club leaders, event organizers, etc. – probably they are included because they can easily and rapidly contact and communicate with many team members. In addition, D2B consumers are in contact with all the "higher ups", and so can quickly get approval or denial from those above them.

One method is just to ask people directly: "Do you get direct-to-brain windows?" and then give them a few seconds to answer. Believe it or not, in my experience, some D2B consumers will actually admit that, "yes", they do get direct-to-brain windows. I can say that at least four people I have asked verbally admitted it. It helps to ask, but be careful not to get into trouble with your job, family, or friends, etc. Once, I was confronted by a supervisor because I told a coworker that some people might be able to hear thoughts. Knowing if a person gets D2B, I think, is the first and probably most important thing an excluded should want to know about anybody, because it makes a tremendous difference. As an excluded, if you know that somebody gets D2B, then you know that you don't need to tell them anything, because they already know it all. Asking if a person gets D2B is mostly a win-win question, because if they do get D2B, then they already know everything about you, there is no harm done, and you are helping to make talk about D2BW more common and acceptable, but if they don't get D2B, then you are telling them about something very valuable, that they may think about for the rest of their lives, and that may help them tremendously in defending against those remotely writing on their neurons. A more subtle way of asking is "Have you ever heard of direct-to-brain windows?"- then you don't sound as crazy if they are actually excluded. It's good to hear what they say, because that represents part of the current public view of D2BW.

Those who get D2BW prefer not to talk about D2BW and RNRAW, while those who are excluded show little or no fear in talking about it.

There are other common characteristics of included and excluded. If there is no clear shilling, and no admission of "I see", then you are left with a person that is either an excluded, or a D2B consumer that is too afraid to admit it. Generally, excluded probably are mostly: males, young kids, elderly, childless, liberals, democrats, non-white, obese, smoke, use recreational drugs, drink alcohol excessively, atheist/agnostic/not-religious, or very religious, poor, not beautiful, have small breasts, have some kind of learning or neurological disability, have done unusual things and got locked in a psych hospital (are "crazy"), are homo- or bi-sexual or transgender, are overly honest, are against restrictions on info and secrecy, care too much about other people, or are overly or openly sexual- are "sex offenders" (note that "are violent" doesn't appear to be an issue but could be). It's tough to be sure, one D2B said "there's no rhyme or reason to it", and probably that is true to some extent. Probably a large part of who gets D2B and who doesn't is simply determined by if a person's parents are included. There may be lineages of families where some have been D2B consumers and the others D2B excluded for hundreds of years.

One tricky aspect is that excluded are routinely made to "mimic" included people. Because anybody can have their neurons remotely written on, an excluded may "sound" like an included by 1) echoing your thought audio, 2) by saying something that makes them seem like they saw something you did inside your house, 3) by apparently "shilling". If a person says a shill that is extremely fast – too fast to be thought out – it was probably involuntary remote neuron writing. It's very tough to quickly replace some work like "think" with "fink" for example, although many insiders know these phrases, get the words tele-prompted in front of their eyes, and can quickly get the shill money window.

Another possible test to see if a person is excluded is to ask them if they know that 9/11 was a controlled demolition. If shocked, embarrassed, and very doubtful, probably they are excluded, but if they already know or actively provide arguments against it, probably they are D2B consumers.

If they have reproduced, probably they are included. I went to a high-school reunion and most of my friends had kids that were about 5-7 years old, and so it's seems likely that most of them got included around 5-7 years before that reunion, which would put their D2B inclusion date around age 30.

Many D2B consumers say hints that reveal helpful info about who is excluded and who is included, for example some have used the phrase "flat out" which implies that females with small breasts (a "flat chest") may more often be excluded, while women with large breasts may more often get included, in particular at the request of many of the wealthy D2B "johns".

Probably many non-white people are excluded. Every time there is a murder that makes it to the news, probably the murderers and victims were excluded non-white people. The murderers are most likely poor non-white excluded people who followed some remote neuron written suggestion to do violence.

A very tricky aspect of D2B is that most D2B consumers, apparently, get a very limited service. For example, you can't presume that just because a person was the victim of a crime, even a violent crime like murder, that they were a D2B excluded. For example, the Kennedy's and many of the victims of 9/11 were probably D2B consumers who didn't get the D2B windows that they needed to see in order to protect themselves from their attackers who many neuron owners and probably even many D2B consumers certainly knew about.

Probably most excluded people are poor people. However, I know of many poor people who do get D2BW, in particular grocery cashiers, because many of them routinely take the money to shill. But the D2B service for poor people must be very limited, because everything costs money - that's why so many poor people shill. Most wealthy people probably don't shill because they don't need the money and it makes them look cheap and low-brow.

Probably many atheists are excluded. In the USA, atheists, ironically for so modern a nation, are less popular that almost all other groups. It seems likely that remote neuron writing is mostly owned and controlled by those who believe in one or more Gods, and in particular (in America, Australia, and Europe) "followers" of Jesus. Perhaps the majority of those who are D2B consumers were "gotten" (included into D2BW) through people in their church, mosque, synagogue, or temple. D2B owners are probably most comfortable with people who follow the herd. My mom was openly atheist and she was excluded all of her life. Speaking from my own experience as the child of an atheist who later became an atheist: if you are an atheist male, put your thumb and first finger together for your daily masturbation, and get used to it, because that's what the Church has planned for you, and they use every dishonest trick in the book, including remote neuron writing, and violence, to do it. That's also about as close to seeing a direct-to-brain window as you will probably ever come. Simply put, those people tricked by the lies and supernatural claims of the religions who own and control most of D2BW and RNRAW don't want the enemy, those who tell the truth, and are supporters of science and atheism, to make more little heretics.

In the USA, under the current Bush legacy (where the Kennedy's, MLK, John Lennon, and other popular democratic leaders are all murdered) probably Democrats, progressives, and liberals tend to be excluded. It seems likely that the majority of D2B is probably Republican owned. They are in the dominant position and don't want to help people on the other side who might help to bring them down, or get them jailed for their involvement in violent crimes.

Teachers, engineers, and doctors tend to be excluded. My dad is a doctor and he was excluded for decades. Teachers are the focus of many sex-with-minors investigations which implies that they are excluded, because if they were D2B consumers, they would know better than to try to sexually touch a person under the age of 18- and have plenty of opportunities for physical pleasure with the many other D2B consumers at the drop of a thought. I'm an engineer and I am excluded. Probably many architects and engineers in the group: "Architects and Engineers for 9/11 truth", are excluded, because not many D2B consumers (who all know that 9/11 was controlled demolition) would feel brave enough to publicly say so and risk losing their D2B service because of violating the code of secrecy. It seems pretty clear that, as was the case for the hording by Pol Pot and a few others in Cambodia, educated people are the targets of the crude and uneducated D2B owners and consumers who remotely write on them, and who haven't done a lot of work in their life.

It may be that excluded do a lot of work, like worker-bees or worker-ants, so perhaps those with tremendous working achievements were excluded. Included D2B owners and consumers, find themselves in a role similar to a queen bee. The argument may be: "we don't want to interrupt their important work by including them", "we don't want to slow or stop their contribution to progress for the excluded and humanity" (as opposed to doing something good for humanity as D2B owners and consumers themselves like going public with remote neuron reading and writing technology). Many D2B consumers must think "maybe the excluded is smart enough to figure it out, and tell the others – if we keep them excluded they can help other excluded people by saying what we can't say", and "maybe the smart engineering excluded can duplicate some of this secret technology and therefore make it public so everybody can enjoy it publicly". Certainly, when a person gets included, they probably instantly get dates, get to make a family, and as a result, their output of non-family related work probably is greatly reduced. But the truth is that the entire system of purposely excluding half or more of the planet from a regular direct-to-brain windows interface is simply a vicious and unnecessary system to continue.

I often walk around and look at all the people around me and wonder who has D2B windows and who doesn't. It must be a sad feeling for an included person to look at a person or group of people and see that they are "excluded"; without any D2B windows, and with only their two "eyes" and "thought" windows above them. Even as an excluded it shocks me to realize that some of the people around me are looking at D2B windows and that others have never even heard of direct-to-brain windows. It must be sad to see people who are without D2B windows, because you know instantly that they are going to have tough lives – constantly misled by remote neuron writing, denied jobs, any physical pleasure and sex, friends, etc. But, like so many classic examples of history, probably many D2B consumers get familiar with seeing the many excluded people, and become desensitized to this senseless segregation.

So that is an important question of our time: "Who is excluded and who is included?" As a subset of that question there is the question: "Who is excluded that knows that they are excluded?". Of the famous people of the past: were they excluded or did they actually receive direct-to-brain windows and pretend that they didn't? So far, I know of no famous person of history who, if they knew that they were excluded, ever left any explicit public record of it. The closest is perhaps William Crookes who said such a thing may be possible. Was Crookes a super-smart excluded or an extremely compassionate-included? Another is Hugo Gernsbach, who also wrote about recording thought-audio in 1911. Was Gernsbach a brilliant excluded or overly honest included? Who knows how many of the great thinkers of the past were excluded from D2B windows? Judging just from people who did not reproduce, perhaps even famous humans like Isaac Newton, and Beethoven were excluded all of their lives. It seems likely that victims of murder and other violence were certainly mostly, if not completely, excluded, or else they would have received a warning in their D2B windows. Even popular celebrities like John Lennon, RFK, MLK, JFK, Nicole Simpson, Gabrielle Giffords, the murdered kids like Larry King, Chelsea King, Adam Walsh, Jessica Lundsford, Jon Benet Ramsey, those kids recently murdered in Norway and their parents, all obviously didn't see the D2B videos of their attacker and their thought screen- which many others must have seen and been watching before, during, and after the violence. One shocking truth that no excluded can deny is that serial killers are allowed to continue their murdering over many years, even when all the D2B owners, and many millions of D2B consumers must definitely and positively know who they are from even simple street cameras, if not from the killer's memories which regularly appear on their thought-screens. Not only did their victims and the loved ones of their victims probably not see the live image and thoughts of the murderers, but they probably don't get to see the live image or thought-screens of the people who wrote to the neurons of the murderers which possibly influenced and/or suggested the violent crimes that were eventually committed.

From my own experience and the many clear examples of history it is pretty clear that the rule has been for centuries that some of the worst and grossest people get D2B while many of the most decent and wonderful people are excluded, and this is a reflection of the reign of religious barbarism. Murderers, torturers, remote molesters with thousands of counts, liars, all get to see, while their gentle and honest victims don't get to see, and are constantly remotely abused; just absolutely upside down.

The top 4 daily reminders that a massive and highly developed secret remote neuron technology exists
4) Being remotely made to itch
3) Having a muscle remotely moved
2) A D2B consumer shills
1) Hearing remotely written sounds

How many humans get D2BW and how many are excluded?
This is another classic question. As excluded, of course, we have to guess, and hope D2B consumers will hint to us. One thing that is clear is that there are many people who get D2B and many that do not. Perhaps the split is 50/50. But probably,

given the difference between developed and undeveloped nations, the ratio is probably more like 3:10, 3 of 10 humans receive D2BW. Many are included, because, in my experience, in Southern California, the vast majority shill in some form or another- those that do not shill are extremely rare- maybe 1 in 30. But yet, clearly many people are excluded because they fill movie theaters, buy newspapers and magazines, books, and watch television. It's tough to know if these "book buyers", etc. are D2B consumers that are somehow doing obsolete things. In addition, in order to communicate D2B consumers are absolutely required by the D2B owners to use the old technologies (like talking out loud, telephones, email, etc.). For example, I see thousands of people, all the time, talking on cellphones, but most of them are probably D2B consumers talking to other D2B consumers, even though a cellphone is completely unnecessary for D2B communication. It's tough to know for sure. One kind D2B consumer said to me "there's a lot out there", which I think is probably very true.

What about the young children of D2B consumers, do they get D2B too?
I don't know and maybe everybody is different – even some pets may get D2B – and be able to communicate through thought-sounds and images. One person hinted that some thought-audio talking does get sent from parent to child – at least when they are very young- perhaps thought-audio like "are you hungry?", but another D2B parent said flatly "we want to tell our kids". My Mom was definitely excluded from birth to death, and probably her parents (and all earlier ancestors) were also excluded.

"Excluded orphans"
Most D2B parents probably constantly communicate with their D2B kids (strictly through thought) while their kids are growing up, in high

school, in college, and after. But, D2B excluded parents with kids, have to use the Internet, cell phones, etc. to see and communicate with their kids. So, unlike D2B kids, the D2B "excluded orphans", those who are excluded, and whose parents are excluded, are the target of D2B abuse. If something terrible is happening to a child, the D2B excluded cannot quickly put an end to it, and may never know about it. D2B parents get to see life from the eyes of their child and see their child's thought screen and so know what the kid is thinking of, and can quickly guide the child into making good decisions and away from bad ones, but excluded kids, have only D2B writers to "parent" and guide them.

In particular, without any doubt, and from first hand experience, many of these young excluded kids are the target of sexually inappropriate D2B suggestions. Because they are D2B orphans, they are easy targets of the many criminals who buy and sell remote neuron writing transactions. Probably many images sent to excluded, are an effort to try and fulfill the sexual fantasies (as seen on their thought-screen during masturbation or sex) of high-paying D2B customers. This, many times, involves sending excluded people direct-to-thought suggestive images of same-gender touching, of them in opposite gender clothing, of them nude, masturbating, or having intercourse in public, or with other species, etc.

Parents should actively inform their kids about sex, I think, to reach them before the remote neuron writers do. People need to accept publicly and openly that the vast majority of young people (after age 10) masturbate every day. It's unhealthy, neglect, dishonest, and idiocy to deny this fact and keep it a secret. We are all products of sex. The current system of denying and neglecting young people's right to physical pleasure is inhuman and shocking. Pleasure and sex education should be a large part of a young person's life. There is a natural desire in most young people to feel warmth and affection. They should be allowed to kiss and fondle each other with clear and constant boundaries and consent, and with the parents being fully aware and helping to actively guide their child's choices.

In particular excluded families should be vocal and active in advising their kids in terms of dating, kissing, masturbating, remote neuron writing, and the laws concerning sex, because even if you do inform and guide your child into regular and healthy kissing and fondling, and daily masturbated orgasm, the D2B writers are going to shape a large part of what guides an excluded child's daily desired physical pleasure decisions. Acting out some of the sexually inappropriate suggestions (like masturbating in public, doing something physically pleasureful with a child they are baby-sitting, or a pet dog, etc.), even though non-violent and many times natural, might haunt the excluded child for the rest of their life, in particular in the bizarre violence-tolerating, pleasure-forbidding society we live in.

The D2B included have an extremely better opportunity to protect their kids, and can interact with their kids through D2BW every second of the day, while being excluded from D2B is to be an orphan with absent parents, and a favorite prey of wealthy D2B criminal owners and consumers. The 9/11 controlled demolition, JFK, RFK murders and their associated cover-ups that still persist to this day are clear examples that laws, even laws against violence, don't apply to most D2B owners and consumers.

"Pedding", "Perving", "Sex Criming", "Insanitizing", "Thefting", and "Violenting" D2B excluded through remote neuron writing

Wealthy D2B consumers and owners have learned from years of experience, that one very effective way to lower the value of their enemies is to use remote neuron writing to get D2B excluded people to commit crimes and do unpopular things. For the many wealthy religious D2B consumers and owners, the targets are probably often liberals, scientists, and non-religious people. For racist D2B consumers and owners the targets are mostly non-white people and those who are supporters of racial integration. It seems likely too that liberals, scientists, and non-religious people may use remote neuron writing to try and lower the value of popular violent or religious excluded using similar remotely written suggestive images, sounds, muscle moves, and other neuron activations. Much of this writing may even be computer conceived and controlled- being somewhat faster and planning farther into the future than the average human brain can. Every second millions of dollars in remote neuron writing transactions occur with just this effort in mind.

One major goal is to make a person do something sexually inappropriate with another person under the age of 18, for example, by sending the excluded many quick images ("flashing them") with well rendered suggestive images of a hand touching a butt, or a kiss, etc. they can get the excluded to decide to act out that remotely written suggestion- to think "will I kiss that person", or "what would that be like?" – without realizing that these evil remote neuron writers are trying to ruin their career, popularity, and reputation. Even to get a person to think of images of young people during masturbation may be enough to mark a person as a "pedophile" or "sex offender" for life, because many millions of D2B consumers can refer to the thought images of the excluded person as evidence of pedophilia.

There are numerous techniques of remote neuron writing, and this book is only scratching the surface. Another method may be to remotely molest a D2B excluded person to make them angry or violent, for example by making them constantly itch, or sending unpleasant and upsetting sounds to their thought-audio (like their mate having sex with their enemies or friends). An angry person is unattractive, and appears to be a dangerous person. Being angry makes a person look bad and doing violence really lowers the value of an excluded person, because D2B consumers will not want to hire or befriend a person who is easily made violent.

Getting an excluded person to do something that will get them locked in a psychiatric hospital is a very popular form of neuron writing. Just getting a person locked in a psychiatric hospital once is enough to forever label them as insane, unstable, unpredictable and dangerous. It's probably easy to get people to make a public disturbance that gets them locked in a psychiatric hospital. One method is to use remote neuron writing to make an excluded person think that God is ordering them to do unusual things, or that they are wildly famous, like a celebrity, or good looking, or extremely wise, and are soon going to be very wealthy. The excluded doesn't realize that they are better off focusing on making sure they have a place to live, on getting a date, a kiss, making a family- because none of those remote neuron writings ever come true- in particular for excluded people. Some part of the trick is that the excluded think that we live in a high-minded and decent society, but the reality, that they don't realize, is that we are living in a dark age where secret violence, lying, and remote trickery are the rule, and where truth, honesty, and decency are the extreme exception, because of the rise of Christianity, and the 700 year-old secret of RNRAW and D2BW. Even the opposite technique is sometimes used, to make a person think that they are of very low value, that they are viewed as perverted, insane, ugly, obese, stupid, old, etc. and then the excluded may be more easily persuaded to do something desperate or to settle for a very bad deal. Some D2B excluded, who realize that many people watch them may be easily convinced to do unusual things to "please the audience", or "have a laugh" with their invisible one-sided audience.

The D2B consumers and owners know that getting their enemy to be thought of as a "sex offender", "violent", "unstable", or a "thief" is a good way to keep them excluded, to lower their popularity and influence, and to restrict their potential wealth, job potential, and in particular their chances of reproducing and making more enemies.

Lots of D2B consumers constantly view me and other excluded people as "crazy" and "unstable", but in truth, and in my experience, as an excluded you never know what to expect from D2B consumers, they flip-flop all over the place, many times for D2B money- one minute they are friendly then the next minute they are mean- but if there is a consistent mood it's mean and sour- certainly for those who fund the constant nasty shill.

There is no question that many methods of remote neuron writing being used are complex and advanced. Much of the neuron writing is somewhat difficult to explain with words, in particular from my very limited excluded position. Beyond that, ultimately, it's clear that if a remote suggestion isn't enough, remote neuron writing can definitely be used to simply move the human mind and muscles any way that is desired. But isn't it interesting how the current system has some kind of planetary unwritten code where writing quick millisecond suggestions (no matter how violent or terrible) are routine, but long periods of continuous remote neuron writing are apparently very rare. Remotely holding lung muscles for a long duration to suffocate a person may be one exception where remote writing is not interrupted for a few minutes.

D2B thought feedback is very helpful

Instantly seeing the thought screen and hearing the thought-audio of a person allows a D2B consumer to quickly see where they may be going wrong, and to adapt. You know exactly what the person you are trying to befriend or date wants, likes, hates, etc. For example, you know they love some particular musical group so you can buy them tickets to see that group, or they like some restaurant, so you can invite them there. You know that they expect you to wear formal or casual clothes, etc. while the excluded knows none of these things. Just like the panty-liner ad on the timeline, the D2B consumer can see a picture of the potential mate kissing them on their thought-screen and know that they are thinking about a kiss – while an excluded would have no idea at all. Also as a D2B consumer, you may know if a person around you is violent, has a weapon, has a violent past, or doesn't like you and you can avoid them. But excluded people never know if a person near them is stalking them, has a weapon, is planning to do violence to them, or hates them; an excluded person is constantly walking through a "mine field". In particular a D2B consumer can use their D2B advantage to know where an excluded person will be, and so can more easily stalk, threaten and attack them. In late 2004, some D2B consumers put dead rats on the bike path I regularly used and near my driveway (fig. 5.3). Some D2B consumers recently exploded firecrackers around me, and I'm sure, often do many other such activities that require knowledge of where excluded people are located that they can only get from the D2B network. It's like something right out of Nazi Germany, but far more subtle, hidden, and protected by secrecy. The Safeway Giffords shooting, the Kennedy killings, 9/11, 7/7, the people recently murdered in Norway, and similar kinds of murders show clearly that many D2B consumers are unaware of violent people around them, presuming that some of those victims were and/or are D2B consumers, which seems very likely.

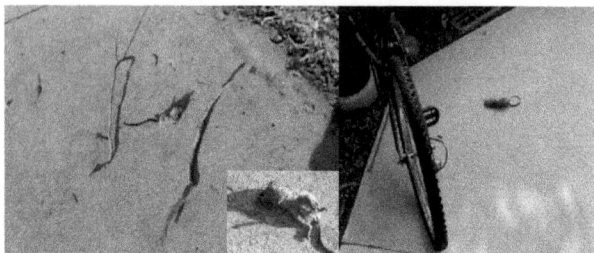

Figure 5.3 Left: A dead rat in an "H" shape on 10/16/2004 with a close-up. Right: A dead rat on the sidewalk next to my apartment on 12/12/2004, just before and after the US election. Coincidence, or stalking and threats of violence by people using a D2BW advantage?

The phony D2B tricks

So much of life on Earth under the D2B secret is phony. Not only do D2B consumers have to talk out loud and pretend that they don't see and hear thoughts during meetings, and even to other family members, but there are millions of other examples of the constant fakery. Here are some:

- Televised "spelling bees" are all fake: only people that receive D2BW can be contestants, and what they say is all scripted and sent to their eyes using D2BW.
- Television "game shows" are all fake: all contestants are required to be D2B consumers, probably because the show owners prefer to have a lot of control over the content and to try and maximize the number of viewers based on audience preferences.
- We have to accept that any sporting event must be subject to remote neuron writing and the free and fast money market transactions of D2BW. People can easily be made to drop the ball, etc. with remote neuron writing. RNRAW has ruined sports in my view because now sporting matches are more a contest between remote particle device owners than between skilled athletes.

Even without RNRAW, I find watching sports dull and meaningless, in particular knowing that our lives are already so pleasureless, and that time is so precious, because life is so short.

- Most wealthy actors, game show hosts and contestants, newscasters, musicians, and teachers, must get D2BW teleprompting to make television shows, movies, concerts, lectures, etc. look more professional; without people forgetting their lines or saying unscripted things. Since D2BW teleprompting exists and excluded people cannot tell the difference, it seems obvious that it would be used all over the Earth.

The Hutu and Tutsi, "second class" citizen nature of D2B included and excluded

Why is it so hard to get a date or make friends? Why do I never get the job? Why are so many people so rude and mean to me? One reason is because not only do those who get D2B not want to be involved with a person who does not get D2B, but because they are paid to discriminate and say rude things. There are a variety of reasons why D2B included avoid excluded:

1. Too difficult to communicate – they have to repeat their thought audio and try to describe their thought-images out loud.
2. Excluded are often tricked by remote neuron writing. The excluded can't possibly know a lot of important information that the D2B consumers know and so a D2B excluded seems dangerous to a D2B consumer. To a human who gets D2B, the excluded, who have never even heard of D2B, are like loaded guns or time-bombs walking around – the sophisticated computers of the neuron know exactly how to play them into a way that is violent or unpleasant to other people. To make matters worse, there is absolutely

no way for the D2B consumer to give the excluded even a single warning like "Look out! That image was remotely sent to your brain...and the neuron writers are trying to trick you!". The D2B consumer must just remain silent and watch, and they may have seen, over the years, much of the abuse and violence done to excluded people by those with remote neuron writers.

3. In terms of a relationship it's like a person that can see and hear dating a person that is deaf and blind. Perhaps these "Ice Castles" stories happen, but it is probably extremely rare. Maybe that "blind and deaf" excluded person is really attractive, but a serious relationship would be too much of a headache for somebody who can see and hear. Mostly this is the case for males, because pretty females are probably quickly included by wealthy included males. There are many better money and career offers, and simply many more choices for those who get D2BW.

4. Being nice to an excluded, makes the D2B consumer look like an outsider. This phenomenon is a classic throughout history. Nobody will defend the accused heretic, because they don't want to be accused of heresy. Other classic examples are: a person defending someone accused of witchcraft, a German person under the Nazis who was sympathetic to a Jewish or homosexual person, a white person friendly to a non-white person during segregation, etc. It simply looks better if you don't have any excluded friends that might call you or somehow intrude in your life and the lives of your D2B friends. All the basics of segregation completely apply.

5. Talking for extended periods of time with an excluded, who knows that they are

excluded, and who talks openly about D2B – as is the case for me – makes the D2B consumer attract labels of "rat" from other D2B consumers and their D2B dealer.

The preferential hiring of D2B consumers over excluded is a massive and terrible truth. Many companies must only hire D2B consumers exclusively – there is not a single excluded person on their payroll. Imagine receiving D2B windows, seeing and hearing the thoughts of everybody around you – of course, you would not want to hire somebody who not only can't hear your thoughts, but is being actively manipulated by very violent and evil people all the time through remote neuron writing – and there is nothing you are allowed to say to the excluded to help to stop it- the D2B consumers can absolutely not help the excluded whatsoever or risk losing their D2B service potentially for the rest of their lives, or worse. The situation for D2B consumers is analogous to the "sunder commando" prisoners of the Nazi death camps, who could not tell each new group of prisoners that they would soon be killed with poisonous gas. In particular, in small businesses, where every penny counts, you don't want the risk of being a humanitarian and hiring a person with only 2 windows over their head and none in front of their eyes (a D2B excluded), because any mistake might end your business.

The excluded are the puppets of the included. Excluded people are used as pawns to sexually arouse D2B consumers. Trying to get excluded to do sexual inappropriate things generates a lot of money for the phone company and neuron owners. Simply put, D2B consumers pay the phone company a lot of money to see sex and violent videos, and this is a big incentive for the phone company to create sexy and violent videos – and many excluded fill the role of actors and actresses because the remote neuron writer never has to face their remote excluded victims or pay for their remote neuron writing.

Certainly one way to stop excluded from doing violent and sex crimes is to show and tell them about remote neuron reading and writing, dust-sized cameras, and direct-to-brain windows. Knowing that some violent or sex criminal might be writing to their neurons might make them think twice about acting on a remotely written violent or sex crime suggestion. Even simply letting excluded people know that their every activity is recorded with microscopic cameras by the governments and telecoms, and can be seen by many thousands of people, would help tremendously to stop many sex and violent crimes. In particular, I think excluded would not bite on illegal and unethical suggestions when they know that the captured images and sounds of them, although currently still secret, might someday be used as evidence to punish or jail them for the crime in the future.

In terms of the Hutu and Tutsi analogy, generally, using remote neuron writing, excluded males are feminized, and excluded females are masculinized. The movie title "Tootsie" is probably a double meaning play on the Hutu and Tutsi apartheid of Africa, in this case the "Tutsi" are the excluded whose males are feminized, for example with remotely written suggestive images of them wearing women's clothing. Other excluded males around them are sent similar suggestions to arouse them into potential male-male encounters, all through remote neuron writing and to the great pleasure and arousal of the Hutu D2B included males and females. Making your honest enemy extinct, by denial of D2B service, in this sense, is very erotic for the D2B owners and consumers. For an excluded male, images of them wearing a bra, women's underwear, touching another male, etc. are constantly beamed onto their thought-screen, in addition to activating neurons that cause the feeling of sexual arousal. For an excluded female, images of them wearing a penis or men's clothing, touching another woman, etc. are constantly written to their neurons, probably like the male, while pulsing the neurons that make humans feel sexual arousal. Excluded people, already deprived of and desperate for physical affection, simply view these thoughts as natural, not as being sent by other people, and are much more receptive to acting out the remotely written suggestive images. This fits well with the D2B desires. The neuron uses the beam to make the less interesting, less profitable, and possibly reproductive heterosexual encounters impossible for excluded.

This is why, although it feels embarrassing, it is good to talk openly about the D2BW segregation, because your own right to reproduce and make a family is being denied, in addition to many other basic rights.

The "Ideological Genocide" aspect of the Neuron lie

Denying millions of people the ability to see, hear and send images and sounds using thought is a massive and extremely effective breeding program that seeks to make the honest, poor, non-white, non-religious, and other groups of people extinct. The lie breeds true. Clearly those who lie have succeeded for many years, not only those who lie and suppress the truth about remote neuron reading and writing, but of course, over an even longer period of time, those who continue the lies and mistaken beliefs of the religions. Those who have no trouble lying about remote neuron reading and writing are readily included, get lots of sex, and reproduce making lots of offspring, while those who tell the truth- about D2B, 9/11, the JFK and RFK murderers, or even the truth about science and religions have to live lives of celibacy and abstinence, not by choice of course, but because it is very difficult to find a mate when you can't see or hear thoughts, and remote neuron writing is in the hands of a bunch of murderers, assaulters, molesters and liars.

First, getting sex for D2B consumers is extremely easy. Only excluded have to masturbate. In fact getting sex is so easy for a D2B consumer, that it becomes boring, and many of them have to abuse excluded people, to keep the D2B male erect, and to make their sex more stimulating. The D2B consumers live out their sexual fantasies by inflicting their fantasies to be acted out by excluded through remote neuron suggestions, of which perhaps 70% are acted on by excluded, while only perhaps 15% of remote neuron written suggestions are acted on by D2B included or excluded who know about D2B writing. Once you know about the existence of remote neuron writing, and this massive "denial-of-service genocide" and "extermination" through hording of neuron reading and writing, that you are on a stage with many voyeurs watching, you learn quickly to reject the suggestions of the neuron writers and focus on trying to get a date, to make a family, etc. But even so, with remote neuron writing in such evil hands, an excluded person even getting a kiss with an included or with another excluded must be extremely rare.

Many times I am sent the thought-audio of a woodpecker pecking, and I think what this means is that D2B women are constantly bombarded with D2B video invitations and offers from D2B guys, some offers, are probably for lots of money for little work in return. Of course, males who are being denied D2BW can't peck on any females and so lose out on any chance of getting a date, a kiss, and ultimately to reproduce. D2B females probably immediately adapt to the money for pleasure system, and/or the pecks of the many D2B males; an excluded male, no matter how handsome, smart, and talented doesn't stand a chance against the many males that can use D2B to communicate with both included and excluded females.

Making excluded live lives of celibacy and chastity is erotic to D2B consumers in the same way that they desire veal. Seeing a sexually frustrated celibate never-touched male or female is highly erotic to D2B consumers. Much of D2B consumer sex is built around teasing excluded people by flaunting the chance of sex, but absolutely denying them any chance of even a kiss. The excluded have to masturbate, but the sex cup overspills for the D2B consumers. Where the excluded are starved for even the slightest touch, the included, who see nude people and their thought screens all day, and can get all the sex they want, have to come up with new creative ideas to make the nude human look sexy, and to make sex seem exciting. As I said, non-violent extermination of your honest and lawful enemies by denial of service and imprisonment is actually very erotic for many people, in particular wealthy neuron owners- maximizing their physical pleasure and minimizing the pleasure of the opposite side. Imagine how easy it is to simply remotely give a human, in particular a D2B excluded, a "bad feeling" about another excluded. It seems likely that computers are programmed to automatically remotely neuron write to stop any kind of romance between excluded people. Misleading excluded people with remote neuron writing is very easy. It's hard to imagine an excluded person having the neurological strength to reject the natural inclination and memories flooding their thought sound and screen.

Not even warning the excluded about remote neuron reading and writing is all part of the massive "ideological genocide", or "excludocide", and is one of the most shocking and inhuman aspects of the D2BW "invisible" segregation.

D2B discrimination goes beyond gender and race; being D2B included or excluded is much more important than gender or race. D2B consumers generally don't hire excluded people, only excluded people hire other excluded people, so, as a result, excluding a person is to extremely reduce that person's chance of survival, because it lowers their chance of being hired, which lowers their chance of gaining income to buy food, shelter and the things they need to survive and reproduce.

No person who gets D2B is going to reproduce, date, or even befriend a person that does not get D2B. It reveals an unpleasant truth about D2B relationships: that the relationship would not exist without both people getting D2B service, and would probably fall apart if one was suddenly excluded- what a deep relationship eh? Losing D2B is probably analogous to one mate suddenly becoming deaf or blind. The only chance an excluded has to reproduce is to find another excluded, and even then, remote neuron writing is definitely used to cause chaos, and to lower the chances of success.

Hand-held and even microscopic beam weapons were probably first invented centuries ago

It sounds crazy, but yet if many people have been seeing and hearing thought for centuries, certainly there must be portable and microscopic particle weapons that can burn people. This includes the very dangerous hand-held maser (like those on "Star Trek" and "Star Wars" but invisible to animals). Many people are not even aware that a CO_2 laser, which is public information, can already cut through metal, and so certainly can cut through muscle and bone. The myth is that a maser (and laser) of high power requires a large device, but it seems likely that not only a hand-held device that can cut a person in milliseconds must have existed centuries ago, but that even microscopic devices may be able to seriously burn and cut living objects. I have felt tiny pin-point burning feelings on my body before and perhaps these are from microscopic laser devices. These devices may be used by two large sides to keep each other in "check"- one side attacking until the burning delivered in response is too intense. Imagine the strongest possible handheld lasers in the hands of average people, for example an old person accidentally slicing off their foot or their neighbor's hand in an instant. But believe it or not, the hand-held laser, like the ballistic gun, is mostly obsolete, surpassed by the far faster, computer controlled, microscopic particle devices which are always armed and pointed at most people every second of our lives.

The importance of "particle supremacy"

"Particle supremacy" is used to hold a monopoly on particle technology. Simply put, if somebody, a brilliant excluded, or even brilliant D2B consumer, tries to duplicate remote neuron reading and/or writing, all of the already existing nano-devices and sophisticated particle device infrastructure can be used to easily disrupt and destroy that effort without any chance of detection. Beyond that, all publishing is highly controlled by the all-powerful neuron owners, and since all major companies are owned by D2B consumer "hostages", they all must bow to the neuron owners, as strange as that may sound.

Without a doubt, micro and nano sized particle weapons are the top of the food chain of most dangerous and powerful weapons. Tanks, bullet guns, robots, missiles, macro air ships, are all mostly obsolete as weapons, and not worth all the tax money spent on them, because they are all penetrable by tiny particle devices. Larger ballistic guns could possibly have value in breaking through thick barriers for the side that has particle supremacy. Any next "World War" 3, as may have been the case for the other two World Wars, might be fought and decisively won in milliseconds. The first and most important war is the nano and microwar, since those devices can easily destroy any larger devices, and are the only devices that can stop each other. But it's not clear, and I am excluded, so I can't be sure.

The light particle, or "photon" is the most dangerous projectile, and particle weapons are easily the most dangerous and deadly weapon known on Earth, and perhaps even in the Universe. Light particles move much faster than a bullet can, and can be sent invisibly from tiny microscopic devices.

Remote controlled microscopic particle devices created more opportunity for wealthy violent people to do remote violent crime and never be caught. The remote nanometer scale particle device is almost the perfect murder weapon, because it is 1) invisible: nobody can see a device firing light particles with x-ray and other invisible frequencies. Microscopic particle devices are very difficult weapons to destroy, because you have to pin-point the source of the particles, which could be a variety of tiny dust-sized hovering devices. 2) Leaves no trace: simply holding a lung muscle so a person cannot breathe for about 3 minutes is enough to remotely and invisibly kill a human or any other species with a lung, and all any coroner or loved one can say is "death by natural or unknown cause"; there is no wound or bruising. 3) Can be remotely controlled: the murderer, assaulter or molester doesn't need to be physically near the victim at all. Since the dust-sized particle devices form a chain from attacker to victim, a person in some underground bunker on one side of the planet can remotely make the nose itch of another person on the other side. Remote controlled nano particle weapons opened up new opportunities to remotely torture people. This is used on me, and I'm sure on many of you, all the time – my muscles are moved, many times, a right finger will be moved, just as I am falling asleep to do "sleep deprivation", your muscles are moved to make you "accidentally" drop things (many times into a toilet or onto a bathroom floor), to "accidentally" cut a finger when cutting food, to "accidentally" burn your finger or arm on an oven grill when putting in or removing food, given a headache, or given a remote "mystery" skin burn or cut.

Throughout history many murderers have adopted a "right" or "left" theme to create conformity and unthinking obedience. Judging from the many murders that we *do know about* done by the current "conservatives" (in the United States, the

"Republicans") of many liberals and popular pro-democracy leaders like the Kennedy's, MLK, John Lennon, and mass murders like 9/11/01, 7/7/05, Utøya Norway, etc., this modern group of evil, like the Nazis, has flourished using the "right" theme to infect the planet with their violence and destruction like a deadly virus. They infect the governments and major media, and just like a violent virus, harness poor, and otherwise nonviolent people in the army and the public to kill for them with very little choice and without knowing anything about RNRAW and D2BW. There have been terrible un-democratic rogue murderers on the non-religious so-called "left", like LBJ, Stalin and Pol-Pot, and there are violent rogues on both sides. Historically, the conservative "right" has probably produced the most first strike violent leaders like Hitler and the Bushes; people who harness the power of the Church, and continue the worst of the Church's brutal and barbaric traditions of harshly torturing and murdering the very best and brightest minds of Earth for petty and trivial reasons and worse- for telling the truth. Clearly they are not right or left, but simply the "first strike violent" and "big liars". A battle of "sides" or "colors" will, I think, in the long term, always lose to a battle of "ideals". Battles of "true versus false", "fair versus unfair", "stop violence versus start violence", etc. will always win over any battle of simply "left versus right", "blue versus red", "north versus south", "east versus west", etc.

Most if not all D2B owners, and many D2B consumers, do not just sit back and deny millions the ability to see, hear and communicate by thought, but they are funding and undertaking a massive and very active war against truth and against those telling the truth (in particular the truth about remote neuron reading and writing, light as a material particle and the basis of all matter, the expanding universe and relativity theories being wrong, and the truth about violent murders like those done by Frank Fiorini, by Thane Cesar, by the controlled demolitions on 9/11/01, etc.). They don't just sit back and allow free speech and the free flow of information to take it's course, but instead wage a constant and often violent war against the truth and against educating and informing the public.

Methods of remotely killing people with remote controlled microscopic flying particle devices
1) Remotely hold a critical muscle contracted, or prevent it from contracting.
2) Induce cancer (William Rollins demonstrated this on guinea pigs back in the 1800s)
3) Use remote muscle neurons to induce a fall down stairs, off a ledge, car accident- how many car accidents were actually remote murder?
4) Use "voice of God" (neuron writing sounds to thought-audio and ear-audio) to trick excluded into suicide, homicide or assault. Another technique is to get them to do something illegal, like a sex crime (excluded can't get sex, and so this is an easy technique), get them caught, and then remotely murder them in prison which is easier.

Remote Neuron writing used to help reinforce false stereotype
One use of remote neuron writing is clear, and that is to reinforce false stereotypes that the D2B owners want to trick more people into believing. The excuses given for excluding people are usually the same excuses, and they center on "pervert" and "crazy" (notice how "violent" doesn't seem to be a big issue- but it can be if it's linked to sex). Most of the time, these labels are applied to people who are just as normal and sexually aroused as any average person. But like the 9/11 19 hijacker story, the Oswald story, the Jesus rose from the dead and Moses parted the Red Sea story – if a lie is told enough, many people will believe it and any

opposition to the lie is viewed as unorthodox, then any person who supports the truth is viewed as a black or stray sheep. All of remote neuron writing is used to try and reinforce these myths; for pervert: 1) by sending sexual suggestions to excluded, 2) by literally moving their eyes to people's buttocks, breasts, etc., which D2B consumers can see on the excluded person's "eyes" screen, 3) by the nature of being excluded: an excluded has to look if they want to see a good looking person while an included can stealthily just call up a live-image window of the hottie in front of their eyes without appearing to stare- and when I say a window of the hottie, I mean one that shows everything – what they look like nude, all the sex they've ever had, all of their most private thought images, their religious and political views, etc. The neuron owners and developers have taken this natural voyeuristic desire to "people watch", to see people in the nude and having sex, and to hear their conversations, what is traditionally called "perversion", to a surprising extreme.

Labels of "crazy" tend many times to be a cover-story or excuse to try to discredit and defame those telling the truth. Classic examples are those who express doubts about the official major media (television and newspaper company) big lies: that Oswald killed JFK, that Sirhan killed RFK, that 9/11 was not a controlled demolition, that people haven't kept RNRAW secret for centuries, etc. who are quickly dismissed and defamed with abstract words like "wackos", "kooks" and "cranks". Also traditionally, those who express doubts about the ridiculous claims of the religions are often, ironically, labeled and viewed as crazy. In general, people who lie are rewarded, while those who tell the truth, about remote neuron writing, 9/11, religions, etc. are demonized and defamed. Scoundrels and mass murderers like Bush and Cheney are labeled heroic, while the actual heroes, like those telling the truth about 9/11, are labeled crazy. So the non-religious and honest are the first to attract the label of "crazy".

Remote neuron writing is then used to reinforce this stereotype.

First, many of the claimed psychiatric "disorders" are not based on clear, diagnostic, and physiological science, and many times the prescribed "cures" (massive involuntary drugging with experimental drugs, electrocution, and physical restriction) are radical, brutal, and many times illegal responses to petty or misdemeanor nonviolent behavior. Secondly, because of remote neuron writing, many people don't realize that what is being called a psychiatric disorder may actually be more accurately described as "remote neuron abuse of a direct-to-brain-windows excluded". For example, what is being called "manic depression", "neurosis", "psychosis", "paranoia", "unusual behavior" – may in fact have an external source. An excluded is very easy to remotely control since they have absolutely no knowledge of remote neuron reading or writing. They may follow the "voice of God" in their head. Many people might not even be aware that many victims and prisoners of the psychiatric system are honestly and accurately describing remote neuron writing when they state that they "hear voices in their head". Many people recognize the classic claim of "people hear my thoughts" and "people watch me in my house". How vicious are the D2B consumers who know that what all these "crazy patients" are saying is, in large part, true, and is the result of the massive secret of remote neuron reading and writing, but still say nothing, and allow drugging, and other "treatments" to occur. Other classic remote neuron example "symptoms" are: an excluded person mistakes someone's identity because through remote neuron writing an image was remotely overlaid over the face of a stranger for a millisecond- the excluded may run up and say "Fred!...oh wait...I'm sorry...I thought you were my friend Fred". With remote neuron writing an enemy of a D2B consumer can be made to have an unnatural devotion to a celebrity, or somebody who they have never met, or a person that abuses them. Movie and television owners, and individual wealthy people may buy remote neuron writing on to D2B excluded (who have never heard of it), for example, to bulk write their image, or send audio samples of their voice, and other messages to maximize the attention directed at the celebrity they are trying to promote, and therefore to increase their money making. Even the enemies of popular people may remotely send images and sounds to try to scare and bother the popular person by constantly bombarding the thoughts of excluded people with memories, images, and sounds of, or relating, to the popular person to unnaturally create a strong devotion, or alternatively a negative impression in the mind of the excluded person. Another technique is to periodically send the excluded an image of a neighbor or coworker. Then the excluded focus their anger on that neighbor or coworker, and integrate those people in their thoughts into elaborate inaccurate theories. In the absence of any kind of established and public scientific explanation of remote neuron reading and writing, some excluded form theories of their own, to explain how many, or a few particular people around them, can apparently see them in their house and hear their thoughts. For example some may think that their metal fillings allow people to remotely control them, that there are tiny "peep-holes" in the walls of their house, or other inaccurate explanations. Many of the explanations center around the belief that God is communicating with them. Thinking that God is telling you to do something, no matter how embarrassing, violent, or illegal, will cause many excluded to act on unusual suggestions. Many excluded want to show their absolute allegiance to God and so obey the remotely neuron written suggestions. Excluded may go nude in public, or even commit gruesome acts of violence, in order to fulfill the wishes of the supposed God who communicates with them through their thoughts.

One popular technique is to play on sexual jealously in monogamous minded excluded. Excluded are routinely caused to experience suspicions and sent sounds and images that suggest that their mate is having sex with other people. Playing on sexual jealousy is a major part of the criminal remote neuron reading and writing empire. It is one of the easiest ways to make a male (or even a female) violent, and to destroy a relationship. Remotely writing images and sounds that try to trigger sexual jealously is probably one of the top ten most successful methods of getting excluded people to do violence, and videos of violence increase the income for the telecoms which sell the videos to D2B consumers. So watch out for remote neuron written attempts to make you sexually jealous- it's better to keep a cool head and realize that people need to be allowed to have sex with whomever they want to consensually, and that much of what is neuron written are lies. Another technique of trying to create sexual jealousy is to make an excluded think they have a chance with some D2B consumer, then paying the D2B consumer to reject the excluded, and walk past them with some D2B mate. To "tease" the excluded is very erotic to many D2B owners and consumers, but also serves the purpose of reinforcing the addiction and obedience of D2B consumers to the demands of the D2B owners; they know to obey the neuron or else they very easily could become that poor celibate excluded "drone bee". A D2B couple may make themselves more aroused by walking by (individually or together) excluded and then paying AT&T to see the excluded masturbate to thought-images of the D2B consumer to enhance their sex (or they might be paid by wealthy D2B consumers to do the tease). One D2B consumer kindly tipped me off to this saying "they're going out", with the helpful double meaning of "that person has a boyfriend already, don't waste your time", and of how some D2B consumers figuratively and no doubt literally "urinate" on excluded people with this "walk-past" teasing. One positive aspect of this barbaric practice is that the D2B consumer probably finds the excluded person attractive. One good way for an excluded to defuse the tease, is to approach the teaser and talk to them for a few seconds, and perhaps try to get a date, because the excluded will then recognize quickly if the D2B consumer is friendly and might actually date them, or if they have the rude, brutal, Nazi-like elitist "leave me alone", poop on the subhuman excluded, viewpoint so typical of many D2B consumers who get anything they want delivered instantly right to their eyes.

Many excluded may express anger or become angry often because of remote neuron writing. Is it any wonder that an excluded person would have anger directed at direct-to-brain consumers around them who are shilling, that is, getting money to say rude comments like "psych", "gay", "vert", etc.? Excluded simply have not the vaguest idea that the D2B consumer people around them are saying these rude things for money, and that the sound and/or image in their mind is not from God, but is probably from wealthy uneducated and violent people who have a shockingly extreme technological advantage of being able to see, hear and write to their thoughts. The poor excluded never even know that such a thing as remote neuron reading and writing is not only possible, but is highly developed, and is in constant use, and mostly constant abuse. Imagine the so-called psychiatric delusion and problems that will be suddenly "cured" when direct-to-brain windows goes public and everybody can see their thought-images of the past.

Another stereotype that remote neuron writing is used to reinforce is the myth that non-white people, by the nature of their race, are more likely to commit crime. By remotely writing suggestions on the neurons of non-white excluded people to do crimes, remote neuron writing is used to trick them into committing crimes. Then the news stories are filled with stories of crimes where non-white excluded people acted on the criminal suggestions and committed crimes, which reinforces this inaccurate belief.

For many years bad people have been the most powerful on Earth and will be for some time into the future

Clearly evil violent dishonest people are at the top and have been for a long time- certainly since the rise of Christianity in the 300s but probably even before then. That is the only way for the murderers of the Inquisition, of the Kennedys, the WTC 9/11 victims, etc. to go unpunished and free, in addition to the many neuron writers who have murdered many men, women, and children, like those kids recently murdered in Norway. The simple fact that direct-to-brain windows has been a secret for an absurdly long time, of the shocking disregard for science (only a tiny minority supporting a theory as obvious as evolution and knowing very little about the history of science), and the shockingly long duration of the belief that Jesus and Mohamed had supernatural powers, etc., are all evidence that, for a long time on Earth, bad has been at top in control, and good has been on the bottom. Whoever shapes and controls much of the planet, the most wealthy, clearly have a violent and dishonest majority view– people are murdered, and their killers are never even identified, massive lies like the expanding universe and 9/11 19-hijacker story continue to be believed for decades, millions are denied even knowing about seeing and hearing thought, let alone regularly receiving it, etc. – just terrible values.

Everything runs on money

One good way of looking at life on Earth is that absolutely everything runs on money. Almost everything that is done, is done for money. With the development of remote neuron reading and writing, and direct-to-brain windows, came an extremely rapid free market. Now, with D2B, many people can receive and accept offers for money directly in front of their eyes. Money is the fuel of people, and many people will only do something for money. But the myth is that famous people are famous because they have special talents. Some amount of skill and popularity certainly plays a role, in particular in sports, but much of the success depends purely on money. For example, if you want to be on television or radio, you have to pay the owners for that time, which is very expensive – but this transaction is, apparently, all done secretly through D2BW. Just like this book – I have to pay to have it printed, and pay to advertise it.

There are entire secret industries that arose with the invention of D2BW. The "shill" is an industry where D2B consumers receive thousands of dollars to say annoying, upsetting and misleading things to those excluded from D2B around them. And that shill industry extends to D2B consumers – people can pay to give them a bad dream, and annoying D2B memories, muscle moves, sounds, images, etc. Newspapers, radio and television news are perhaps the most misleading – Fox news, is a fine example, but they all exist strictly from payments of money to deliver information that the buyer gives them. For Fox news it's basically the 9/11 murderer group who fund so much of what is told to the public as if it was actual news. Notice, for example, how no major news source ever reported the discovery of D2B, the truth about Frank Fiorini, Thane Cesar, 9/11, 7/7, etc… or even that anybody has any doubts of the Republican and Democratic representative and major media company's version, or even that there *is* a second version or explanation. Mostly, the major media are advertising companies – and the most convincing ads are disguised as "news"- all of which is paid for and purchased by D2B consumers and owners through the D2B silent and invisible (at least to excluded) interface. Think of the D2B consumers who turn down thousands of dollars to shill and do other unpleasant things. It must be very difficult to not accept an offer for $5,000 just to say "gay", or $500 just to walk by some excluded person and cover your mouth with your hand. Perhaps the view of those D2B consumers who pass on all the money is that there is a higher potential future money for not shilling or taking money to do immoral or unethical things.

It is simple to understand that all the television, radio and news companies survive from large payments for advertising, and I think when D2B goes public, the public will finally see that probably every single image, sound and sentence broadcast and published by the big media companies is a paid-for ad; many times paid for secretly, using the "absolutely no-paper-copy" D2B network. This is obvious when you realize that, even just for here in the United States, all of these terribly obvious violent murderers are on the loose- those who did the 9/11 controlled demolition, the killers of JFK, MLK, RFK etc., but the major stories being funded, printed and broadcast to millions of excluded people are mostly about trivial non-violent activities. It's clear that one goal of those who fund and own these companies is to change the focus of the excluded people from serious violent crimes to a more anti-pleasure, anti-drug, anti-free-info focus. For example, in the interest of survival, the clear focus in my mind is to show and tell the truth about remote neuron reading and writing, evolution, the history of science, and our future as a star cluster, about full democracy, total free info, violence as the big evil, pleasure as something to consensually teach, and openly celebrate, etc. In particular a clear important focus in my mind is that the public should see all the crimes of the worst of the violent, to remove the worst of the violent from remote neuron writing, and from being able to do more violence. But showing the truth about 9/11, the Kennedy killings, the galvanizations, etc. and exposing those murderers who live free and unseen on the loose is not the main story, instead, accusative stories without any actual physical evidence about inappropriate non-violent sexual activity (in particular with children) are the main stories. To their credit there are stories about poor (presumably D2B excluded) people who do violent crimes, with low-tech fists, knives and guns, etc. but then always the explanation is not remote neuron writing puppetry, but the immediate application of unlikely psychiatric theories. Even with the remote neuron writing "television" (the unconsensual one on your thought-screen and in your thought-audio that you can't turn off or block), the views paid-for and written are always that sex and sanity are the focus and topic of everything; violence and exposing widely believed lies is never a concern. The "neuron" is so powerful, in particular with their remote neuron writing on the "sensitive" brains of excluded people, that there is a strong argument that we can read any news story by replacing the names of the people with "neuron", for example, we can replace "the suspect did this" with "the Neuron did this". When a person murders, we must ask "why did the Neuron murder?", when something is stolen "why did the Neuron want this stolen?", etc.

There is probably an unavoidable and somewhat brutal natural economic and biological truth in the deciding by the majority of D2B owners and consumers which humans may continue to live and which may die. It may even be that the main reason some person is still alive is because they are doing something (telling people about RNRAW, 9/11, helping, etc.) or have some kind of value (are pretty, funny, etc.) that justifies the cost (in money and even perhaps in lives) to protect them from the many violent people on the other side who remotely end the lives of many people (who, it may be, do not justify the cost to protect them) all the time.

There is a funny video from SCTV, "Dan Money"[124], where Joe Flaherty has to pay everybody to

---

[124] "Dan Money", SCTV, 11/28/1977
http://www.youtube.com/watch?v=jvEGOEA0OiA

cooperate, for example, the judge asks the jury foreman "has the jury reached it's verdict?" to which the foreman says "well your honor, maybe we have and maybe we haven't...", and Dan Money has to pay the foreman to get the verdict he wants. We can laugh at this funny 1970s television cop show parody, but this skit may be hinting at a darker meaning given what we know about the long-lived secret of D2BW. The truth is that millions of poor D2B consumers march all over with little money windows in their eyes. It seems very likely that direct-to-eyes money offers through D2BW certainly have corrupted the justice and many other systems. Probably many people are called to court or jailed because some people (including members of police, a jury, and/or a judge) were paid money through D2BW.

Violence and sex shows and games

It's gruesome to talk about, but it seems likely that there may be pre-murder "shows" or "videos" that D2B consumers can buy from the telecoms. These videos may have narrators or commentators describing the potential murderer, the thought-images of the murder that the potential murderer is planning, the profile of the potential murderer and the possible victims. These videos are probably advertised in pop-up windows to the D2B consumers who must pay to see them. Just like there are "get money" windows, of course there are probably many more "pay money" windows. In addition to the pre-murder videos, there most certainly must be also be "post-murder" videos of the murder during and after the murder from the eyes of the murderers and victims, and from micro and nano-cameras hovering around the scene, which the Neuron companies sell to the public. These post-murder D2BW videos are probably much more common than pre-murder D2BW videos, because there is no chance of D2B consumers interfering with the murder after it occurred.

In addition there may be sounds of thought-audio cheering, and applauding. When a violent scum is murdered good people probably applaud in their minds, and when a decent person is murdered the criminals applaud in their mind. A classic example is the way US President George Bush jr. clapped at the same time a plane crashed into the World Trade Center on 9/11/2001. Bush sees this plane crashing into the WTC in his D2BW and then he and other like-minded D2B consumers must clap to reaffirm approval of this radical direction that they took.

In addition to the violent videos there must be a big industry of sex videos, and the sex videos of the neuron must be very different from those available to the general excluded public. For example, the D2B videos certainly are of real-life sexual events, many of which are not scripted. Many of the D2B sex videos certainly contain violent crimes. Without doubt many of the D2B sex videos contain people under the age of 18. Probably the videos of excluded people who bite on sexual suggestions are very erotic – in particular because they are so innocent in not knowing anything about D2B, are very sexually frustrated, and don't realize that many people are watching them. But even the videos with D2B consumers may be quite erotic because they are probably much more sexually experienced than excluded people. In addition, D2B consumers can participate in the massive and fast moving money for pleasure market in the D2B, which the excluded cannot access.

In terms of games, there may be gruesome games that revolve around sex and around violence. I hesitate to describe this because it could be unpleasant to some people, but in the interest of knowing the truth about what D2B consumers and owners might be doing, I decided to include it. I can imagine that there may be games like "which one lives?" in which viewers vote to determine which of two humans will be allowed to live. Most of these games probably involve excluded people who are like "throw away" people, and the excluded certainly can't see to complain. Probably "Russian roulette" is played with various death-devices like guns, perhaps with prisoners of war, or with people who agree to take the suicidal gamble for thousands of dollars. Some other gross games might be "snuff the excluded"- trying to use an excluded or D2B consumer with a gun or weapon to hurt and/or kill another excluded (or even a D2B consumer) using remote neuron suggestions. You can see that something like that was probably done to John Lennon. Perhaps wealthy people may use their vast wealth to pay for "irony" sex to occur at the same time they are trying to kill excluded or D2B consumers (for example some poor educated young woman who is gang raped, or takes thousands of dollars to "hand" a bunch of neuron criminals while they try to remotely kill her husband, boyfriend, or a popular enemy of the funders). Possibly the phrase "bang bang" which I just heard in thought-audio, sometimes applies to those kinds of videos. We can only imagine what kind of vicious things the ultra-wealthy D2B owners participate in to make them aroused and entertained.

The sexual aspects of the direct-to-brain segregation

One thing is clear and that is that D2B consumers can easily get sex, and those excluded from D2B mostly cannot get even a kiss, and certainly not sex. I apologize if you are a person who is offended by talk about sex. I prefer to take a scientific view of sex as a natural biological phenomenon on Earth. It is helpful, in the interest of knowing the truth and protecting ourselves, to understand that much of what is happening around us may involve some kind of secret hidden sex. Beyond that, consensual sex and talking about sex is nonviolent, and we need to educate ourselves about the science of sex, anatomy, and physiology.

How sex operates for D2B owners and consumers is not entirely clear and we can only guess about what they do. Clearly many D2B consumers hint about "coatings" and "coverings" and "huddles" which implies that a number of males coat or cover a single female with their sperm. When you can see and hear thought, it is very easy to hook into the system of getting sex. With D2B, the D2B consumer can easily and quickly know who has a sexually transmittable disease (STD), and who doesn't – while excluded can only guess, and probably default to not wanting to take a chance of catching an STD. It may be that excluded and D2B consumers may even be targeted with potential sex partners who have STDs – 9/11 and so many remote murders shows that the neuron owners and many wealthy D2B consumers are vicious, cold-blooded murders and assaulters, so giving excluded enemies an STD would be mild in comparison.

The logic of why the D2B pleasure industry is probably huge is simple, you can either regularly masturbate yourself, or have somebody else masturbate you, you can either think of a woman or man, or actually touch one. Of course, most people prefer to actually touch. The various forms of love, like physical caressing, simply feel good to many humans. Beyond that, when you see many and most people participating in a money-for-pleasure and dates system – most new D2B consumers must happily and readily jump right in, and their lives suddenly become much happier and pleasure-filled. I don't know for sure, but it may be that for many D2B consumers, that males have to pay females for a kiss, a date, etc. has become more of a symbolic formality like tipping.

Clearly there is a massive money for physical pleasure system for D2B consumers. Getting money-box windows and videos direct-to-brain (appearing in front of your eyes as semi-transparent windows or on your thought-screen) is an extremely fast way to make transactions. The D2B consumer simply thinks the audio "yeah", "more", or "no", or even just draws the window closed in their mind, and the computer understands to close the window or "up" the offer.

Put yourself in the place of a poor female or male: they need to pay their rent, have no job, can't live with parents or relatives, and there are the money box windows with easy money – and all you need to do is to just let somebody touch you, or kiss, or rub somebody– for thousands of dollars – enough to pay your rent, for a car, or for a college loan, etc. Once robots are doing all manual labor tasks, one of the only remaining paying jobs for most humans may be physical pleasure for money, so it makes sense to officially and publicly end the prohibition on consensual prostitution for the D2B excluded people too.

One insider hinted that perhaps half of the D2B consumers do pleasure for money, the other half perhaps opting for a more traditional monogamous pleasure-based relationship. Clearly D2B consumers can browse the profiles, directly to their eyes, of many potential mates to make a family with, including seeing each potential mate in the nude, their lifetime accomplishments, and many intimate details of their lives. Excluded people only have the Internet and must choose from those in their immediate surroundings, many of whom are D2B consumers who will not even befriend them.

Some of the more well known aspects of D2B sex rituals are: "see you laid-her", "laid out", and "cuckoo". It seems possible that many D2B consumers can easily seduce excluded people, because they can see their thought-image, hear their thought-audio and deliver them exactly what they expect to hear. So many males, for example, may have sex with a large number of included and excluded females – they may "lay-out". The saddest part is that the excluded females think the insider guy is being monogamous, and this may arouse the D2B male even more- to be tricking or "using" the poor excluded female. The cuckoo is a bird that replaces an egg from the nest of some other species of bird with one of her own eggs, and then flies off. The other birds mistake the cuckoo egg for their own and raise the baby cuckoo. Many women get D2B, because the neuron owners know that women can generate a lot of money for them from the many wealthy direct-to-brain "johns", men who use Ma Bell, or *"Madam"* Bell to browse, shop, and eventually pay D2B women for physical pleasure services. Isn't it a surprise to find that AT&T is perhaps the biggest and most successful "pimp" of history? Much of the money AT&T gets is from the work done by many "hand-maiden" women. There is probably a similar analogy with the many women operators that work for the phone company. Direct-to-brain females must get many many money windows with offers for sexual jobs. It sounds probably very blunt and crude to excluded ears, because the well-oiled D2B pleasure-for-money industry has reduced the system to quick terms, but many transactions may be for "hand", "suck", or "hole". It must be very tempting to accept these easy, fast and STD-free jobs. Many jobs probably provide lots of money for very little work. For example $5,000 dollars just to masturbate a guy, $500 just to let some guy fondle their breasts for 10 minutes, or for a kiss, or $200 to caress their butt for 5 minutes. You can see how an excluded could never compete – they have no idea that you have to offer money, and that not only that, but lots of money. So a D2B female may find an excluded male who she can live with, marry and perhaps even have kids with, while working during the day getting thousands of D2B dollars for D2B sex jobs, even just for "handing" D2B guys, while her excluded husband is at work, and the excluded husband never finds out about it, how could he? But certainly D2B women must get pregnant by D2B guys, and an excluded male raises the kids not even knowing that they are not from him, and so you see the analogy of the "cuckoo" (although with the D2B cuckoo, a male fertilizes the female egg and leaves the child to be raised by a different unknowing excluded male with the original mother).

Much of the "see you laid-her" (which D2B consumers hint about), is that wealthy D2B males will pay a female to walk by some excluded male, to "tease" him, to make him think he might get a date or maybe even sex. Sometimes they may even be paid to make sexual gestures to the excluded. Then they pay this female for sex while they watch the recorded thought-images that the male had of the female when he masturbated. For a female, seeing a male with thought-images of herself during his masturbation is probably a turn on for her. It probably is erotic for D2B consumer male(s) having sex with the D2B female(s) too. It's tough for an excluded not to think about the people around them when masturbating, because that is all they have to choose from, and excluded are starved for even the slightest hint of affection anyway.   One D2B female audio I received said it well (paraphrasing): The excluded gets the trick, the D2B consumer gets the treat. The excluded gets the lie, the D2B consumer gets the lay. It stands to reason that since no D2B consumers ever respond to your thought audio calls of "come here, and let's talk", or "how about a kiss?", or interact openly with your many thought-audio questions and comments, that talking out loud to them would make no difference either. But perhaps the D2B "teaser" might actually "work out", or might respond, but only if you talk to them out loud – it can't be ruled out in theory.   Plus, I don't think people should view "teasing" as negatively as perhaps many might, because, although it is a terribly imbalanced, unfair system with evil people in charge, still, it is likely that "the teasers" are sometimes attracted to the poor excluded target of the paid-for teasing. Being friendly to a person you might not necessarily have sex with is natural, and so in some sense, innocently, and without any mean intention, a person may appear to be teasing, but is only being friendly. For some people courtship takes a long time and there are many tests analogous to spider or swan movement-matching courtship. Some of teasing can be explained as a need for variety, which drives sex for many people. The need for variety in sex is similar to the need for variety in other parts of life; we don't want to eat the same food, watch the same movie, and listen to the same song every day, because it becomes boring after a while. Many times people may have a partner already, but are open to, or actively looking to find a new and better match. Beyond that, the teasing of excluded people is nonviolent; and of course, violence is the big evil, and the thing to fear and to be angry about.

Beyond watching an excluded person masturbate to thought-images of the paid-for "teaser" who walked by them, another aspect of "see you laid her" is that a D2B woman may be aroused by an excluded male (whose image is clearly seen on her thought-screen during masturbation and/or sex), and D2B males may then pay this D2B woman to remotely "abuse" the excluded guy which many D2B owners and consumers might pitch money in to see and may find erotic. There is a double-word score for the evil D2B in that, possibly, if the excluded male was included, he and the D2B woman might mate and make a family together; so they keep both (perhaps like-minded) people from reproducing with this scheme.

Another possible scheme is that a poor D2B woman may be paid a lot of money to have sex with the co-workers, worst enemies, or best friends, of an excluded person who often has images of her on his masturbation screen, or who often appears on her masturbation screen. Even if a D2B woman doesn't take the money, the D2B consumers around excluded people are paid to shill words to that effect by implying that this kind of sex happens, to try to amplify the sexual frustration of the excluded.

Even young post-pubescent kids in high school are the subject of sexual remote neuron writing. The males the D2B helps are many times violent and uneducated. Remote writing is used to help these males get sex, and to prevent the opposition, their enemy, the bright, non-religious (or simply non-Christian) males from getting sex. No matter how handsome they may be, the smart and non-religious males then lose out – not naturally, but all through remote neuron writing with an extreme 9/11, Kennedy killing, Inquisition-like bias. As if by magic, the young females just mysteriously feel aroused by the crude guys, while the nice guys mysterious say all the wrong things, are mysteriously too shy, make all the wrong decisions, and just produce a negative impression in the young women's minds. This exact same, "sexual shaping" using remote neuron writing continues after school too. Part of this system is already openly clear in the choices of who gets D2BW- all the most violent and crudest of men, like the planners of 9/11 and the Kennedy murders. Simply receiving D2B increases the chance of sex and reproduction of the violent and crude males.

Apparently, the more frustrated the excluded feel, the more erotic it is for many ultra-wealthy, decadent D2B owners and consumers. While an excluded may be aroused by something as small as the scent or voice of a female, the D2B consumer can get all the sex they want, and so they have to create new and extreme situations to make sex exciting and interesting in order for them to naturally stay erect and aroused. For example, young attractive women may be paid thousands of dollars to "molest" excluded by remotely making them itch, or moving their muscle by following the commands they see in semi-transparent D2B windows, while some drooling thug holds them by a chain and has paid-for sex with them. I can only imagine what the public might see if D2B ever does go public.

"See you laid her" might also relate to poor women who want to see videos of people in their houses and their latest thoughts, but don't have any money, or even simply because the neuron forbids it, so in order "to see you", they have to do hands, sucks, and holes (ergo "laid 'er"). To see people they want to see, they have to do sexual favors, for example, have sex with the neuron owners or their friends. It is the classic story of "you don't get to see, until you...." and then fill in the blank.

In addition, in the process of a D2B consumer finding a mate, the D2B consumer may entertain the possibility of an excluded getting included at some time, and so send romantic thought-audio and images to the excluded, but all is for not if the excluded never gets included, which is probably the rule, and so it results in a system to the liking of the evil D2B owners and consumers who view such messages as a "tease", since probably 99% of the time, that friendly voice in an excluded head is from somebody they will not only never meet, but probably never even see, and who may not even exist.

Many excluded have very old fashioned sexual views, and so we have to remember that as long as no body is hurt, and there is consent, most sex is just a bunch of "goo"; violence is much worse. An important lesson for excluded people is that sexual jealously is stupid, because if you really care for somebody, you would let them have sex with whoever they want to- you would not try to control them.

It is unpleasant and shocking to write, and I have doubts about even including this in my book, but have decided that it is better to get as full an accounting of what may be going on as possible. It seems likely that there are also many violent videos that the neuron owners (again presumably the telecom companies) make and sell to D2B consumers, for example, "snuff" movies, where, mostly excluded people are murdered- some being murdered during some kind of sex event. Probably poor women are choked to death and the neuron owners sell the videos directly to people's eyes, not only of the scene, but from the view of the eyes of the victims and murderers, and of their thought-screens and thought-audio- which must be very unpleasant and shocking. This may be one aspect of how the D2B consumers, as some have said, apparently "a-pre-she-ate-it"; that is, seeing videos before a poor excluded or D2B consumer is murdered, to emphasize the importance of secrecy. The neuron owners know that D2B consumers pay hundreds of dollars to see such movies. It may be that many people who die were actually secretly murdered in this way. I can imagine a liberal victim getting thousands of dollars to masturbate a bunch of conservative murders, who may then assault, urinate on, and abuse the victim. There is probably an endless supply of poor and hungry humans who have no money or place to stay that do unpleasant things for money. Another classic scene might be that the D2B consumer thinks they are just going to do a "hand", but are then murdered- and nobody is punished because it's all a big secret, "under the neuron".

In summary, it seems likely that a very large and active money-for-pleasure market exists for D2B consumers, and that shockingly and idiotically, and perhaps erotically, they deny the public even an equivalent vastly reduced excluded service (since they work so fervently to keep even asking to be paid for or to pay somebody for pleasure illegal). It's tough to know without seeing the thought-screens of the wealthy and those owners of the telecom companies. Maybe we do live in a very sexually celibate age where even most D2B consumers wed, remain monogamous for life, and any thought or talk of "coverings" and "coatings" is only another big lie or futuristic wish by the D2B owners and consumers.

Common mistaken beliefs by people excluded from D2B

Being like a blind and deaf person, a D2B excluded has many common mistaken views. One mistaken belief is that you are going to tell the D2B consumers something they don't already know. This is a big mistaken belief that, of course, the neuron owners and consumers do nothing to stop. Perhaps there are exceptions, like the details of evolution and the history of science, the truth about religious history, the alternative views to the propaganda they receive, but generally, the D2B consumer already knows just about everything that you might possibly tell them- because, it seems likely that they have access to the thought-screens of many years of history – they already know that all matter is made of light particles, that light is a material particle, that the expanding universe theory is fraud, what you had for breakfast, what you just thought, your mother's maiden name, your social security number, all of your secret passwords and combinations, where you keep your valuables, what your most embarrassing moments have ever been – they know it all already as unbelievable as that is to accept. Talking out loud is basically obsolete for those who can see and hear thought, but may still have some kind of function, and must, of course, always remain legal and unpunished. Mostly, any time you see an image of some person who is a D2B consumer, and think of something you may want to say to that person – like rehearsing for a future conversation – for example thinking "when I see that person, I will say this..." – it's already too late- most D2B consumers apparently review the thought-images and sounds of any excluded people that thought of them (had their image in their thought-window or said their name in thought-audio), so they know long before any future conversation might happen what you plan on saying or doing. In short- you are having that conversation now, or they will probably be aware of that thought, long before you see them in person again. Most conversation for D2B consumers may be only in thought-images, being the fastest method of communicating some idea, but also perhaps in thought-sounds too.

Excluded people constantly forget that the D2B audience already sees and hears their thoughts, and so they say out loud or act out the sounds they hear or see in their thoughts. But for the D2B audience, it is generally annoying, because it's like seeing or hearing the same thing twice- like an echo. Many images on the excluded thought screen are images that are externally sent there, not that are internally written by the excluded brain. Alternatively, a thought image may be externally triggered to be internally written onto the excluded thought screen (by making you remember some image you actually saw earlier). Probably most images of people, in particular of famous people, that are sent to excluded people's thought screen are identity theft-sent by somebody other than who the image is of. This may be just to annoy the poor excluded people with reminders of people they don't like, or to try to get the excluded to dislike the person in the image, or to get the excluded person's opinions of the person in the image, or simply to advertise the person in the image. Possibly some images sent to excluded might be from D2B consumers that are trying to get your current opinions about them. Another possibility is that images sent to excluded people's thought screens, may be to convey some message to an excluded (that is, to use remote neuron reading and writing as a regular telephonic communication system with those who are excluded- which, in my experience is rare). As excluded this makes you realize: how many people have your image on their thought-screen but you don't get to review their thoughts? What important thoughts and news about us are we missing?

Who owns and controls D2B?

So who owns all these microscopic remote control cameras, microphones, and neuron reading and writing devices if such things actually exist? I don't know for sure, but it seems very likely that the telecommunications companies own and operate most, if not all of them. This includes the telephone (both wired and wireless), and cable companies. One argument against this is that neuron reading and writing apparently predates the existence of communications companies (although communication companies may go far back into the past). But clearly there must have been companies formed to sell D2B service quickly after it was invented. The initial discovery of microscopic microphones and cameras must have been very exciting, and after the initial celebration, there must have been a lot of thought about how to distribute the new technology among the inventors and their friends. Probably, long ago in the past, perhaps even in the 1200s, wealthy science-minded people developed camera, microphone and information recording technology and also, in the process, discovered remote neuron reading and writing. Then there must have been a massive focus on miniaturization. Probably they used their money to form small secret businesses to grow the technology, but all secretly- just among their wealthy friends. They do the science, and then they and wealthy investors use their money to fund the labor involved in growing and mass producing the microscopic camera and D2BW technology. Those early inventors and developers of micro-microphones, communicators, cameras, and neuron readers and writers must have enjoyed sharing video and sound recordings with each other (probably many of nude women and sex, but also of secret important truths surrounding crimes, and scientific ideas and technology in their own and other nations), but also it was probably important for them to be in quick communication with each other in case of emergencies.

Another possible RNRAW owner and developer may be governments; in particular the military portions of governments, or perhaps even the communications ("postal", etc.) portions of governments. Perhaps the neuron companies and developers are divided by national and even smaller territorial boundaries.

Whoever it is that owns and controls most of RNRAW and D2BW, they must have all the best answers to any questions humans of Earth might ever ask, because of the many years of secret scientific research that resulted in the many invisible secret nanodevices that we feel the effects of everyday. Unlike their poor excluded victims in the public, the D2BW owners are probably very difficult to trick and deceive; they probably don't waste a second of time entertaining all the far-fetched theories the poor public is bombarded with and wrongly believes (like the claims of religions, that 19 hijackers did 9/11, Oswald killed JFK, the universe is expanding, that time contraction and/or dilation can occur, and that light is a massless electromagnetic wave).

It seems likely that there are at least two major Neuron groups, just like there are two big political parties in every nation – and probably the division is along traditional lines of non-religious vs. religious, conservative versus liberal, majority rule versus minority rule, etc.

Another important question is: Throughout history have communication companies recorded copies of all the messages they delivered? Perhaps telegraph companies secretly kept copies of telegraph messages. They must have known the value of knowing what important people are saying to each other. Once a message is in electronic form, making copies of the message can be done very discretely. Then the secret information can be sold to other people to get more money. It seems very likely to me that the phone companies (and now the cable companies too) have recorded every phone call ever made on all the wires that they own. Wouldn't it be fun to hear famous phone calls? It would be fun to hear our first phone call too. That the phone company probably does record many phone calls and Internet information, implies that they also record thought images and thought sound communications. If true, then the phone companies' archive of data must make the Library of Congress look small. Alternatively, if some part of government is involved, perhaps the actual Library of Congress is much larger than the public realizes, or perhaps there are very massive military data archives that the public, who funded them, does not have access to. It seems very likely that many images of people inside their houses, and their thought images and sounds are recorded and permanently stored by the RNRAW owners and controllers. I, for one, certainly hope that every single neuron writing transaction has been stored, because many of those unpleasant remote neuron writings need to be "paid-back" onto the neurons of the sender.

Have the owners of neuron communication technology and neuron segregation had an undemocratic and unfair influence over the representative-democracy governments of Earth? Certainly by excluding many millions of people from access to D2BW, they are restricting democratic voting, because it's difficult to count the vote of somebody who can't see or hear what is being voted on. Many D2B excluded form very inaccurate views about politicians, because the excluded don't see all the violence that the politicians secretly support, like 9/11, the Kennedy killings, etc. A good example is how, as a 19-year old excluded kid, in 1988, I actually voted for George Bush Senior, not knowing that he was so closely linked to the murder of JFK, and probably the murder of many innocent nonreligious people. If the future of government is full and constant democracy where people vote directly on the laws and decisions they must live under, then you can understand the importance of having a transparent and democratic neuron communication service where the public can see if there is any dishonesty or secret agreements. It must be a natural system for some wealthy people to want to horde resources for themselves and leave the poor as uninformed and uneducated as possible. Hitler and Pol Pot were two of probably many typical examples of popular governmental leaders who may have been hand-picked puppets of RNRAW ultra-wealthy owners that prefer to leave the public in ignorance and to trick them with big lies.

Because all of remote neuron reading and writing has been secret for 700, or however many, years, imagine how many humans who labored to develop and improve the remote neuron reading and writing technology are completely unknown to the public. Of course this list of people also includes many unknown remote particle heroes and villains.

If you think that the public should know about the history of science, about remote neuron reading and writing, and direct-to-brain windows then you are an instant enemy of those people who currently monopolize RNRAW, and they use their extreme technological advantage against you. It's helpful to try to put yourself in their position, and see the universe from their eyes. Each new generation of D2B owners faces the same situation. They are all born on a runaway train of D2B excludocide of much of the human race through denial of providing them with direct-to-brain communications. So you can see why each generation of D2B owners would fear going public with RNRAW and D2BW, not only because it increases the competition for the best resources, but because they might be blamed and punished for centuries of abuses done by generations before them when many people see how they (and those before them) kept RNRAW and D2BW all to themselves. The feeling that the people of Earth would look at them like they are heroes for going public with D2B is probably out-weighed by the fear that they would be viewed as scoundrels for keeping RNRAW and D2BW for themselves for such an extraordinarily long time. But certainly, they must know the public mind and human behavior so much better than we as excluded do. In addition, the current generation of D2B owners ever losing control over RNRAW and D2BW seems unlikely, but as excluded we can't be sure.

Without question the controllers of remote neuron reading and writing technology have very low ethics compared to average people. This is the only explanation that can account for the way the most obvious truths are mysteriously and secretly denied-raising thought-audios like "...gee, was JFK shot from the front?...how could Sirhan cause powder burns on the back of RFK? ... could light be made of material particles? ... might remote neuron reading and writing be a valuable science?... how could three steel frame buildings fall into dust from fire?...could that 1936 red-shift be from the different sizes of the spectra? ...", and so on with the most basic and obvious truths that are forbidden by each successive generation of D2B owners. Because the D2B owners and operators are so terrible, the D2B consumers so addicted and scared, and the D2B excluded so under-informed, these ridiculously simple observations continue to fail to reach the majority.

The picture in my mind that seems to be confirmed all the time, is that the neuron owners, even those on the "good half", are not a group of gentle-men, not intellectuals, not a sensitive people. But instead the view is that they are a pretty crude, low-brow, anti-education, greedy, very selfish, overly-tough, dishonest, group of people that have very obviously slanted and unreasonable views. But, to their credit, I have to admit that I am still alive to type this and you still alive to read this.

Life under the Neuron is very similar to life under a "Cam-munist" and "Not-see" government

There are a lot of similarities between the current system of life "under the Neuron" and life in nations that live under a single life-long all-powerful supreme leader and dictator like North Korea. How much are the neuron owners like a "Kim Jong-il-Neuron" and son "Kim Jong Un-Neuron"? Just like in North Korea, the neuron owners are a singular, all-powerful, life-long, central dictator. Just like medal and award winners thank the "supreme leader Kim Jong-Un", so medal and award winners under the Neuron "appreciate it". Those under the Neuron must be even more humble- not even thanking "supreme leader Neuron" proudly, openly, and vocally, but only indirectly through praise phrases like "I appreciate it" and "keep in mind". Anybody who objects is sent away to the virtual gulag of "excluded" for large periods of their life. Anybody who tells the public the truth about remote neuron reading and writing and direct-to-brain windows is labeled a "rat". Any remote neuron writer who sends an excluded person unauthorized D2B windows is no doubt harshly punished with exclusion, loss of job, loss of D2BW, and probably lots of remote molestation sent by their former fellow remote particle "capos". People cannot even admit openly and publicly that they receive direct-to-brain windows, and that they can see and hear thought. Not only are they strictly forbidden by the supreme leader from admitting that they get D2BW, but cannot even be silent or say "I can't say" if ever asked - they absolutely *must lie* and say "no". Many D2B consumers most certainly must take it even further forcing those who ask if they "get direct-to-brain windows" to undergo counseling and psychiatric detention. D2B consumers are intimidated and coerced to vocally lie about seeing and hearing thought even to their children, parents, and friends. They are often forced, out of the blue, to make a clear "pledge" or "oath" statement out loud by stating "I don't think people can hear thoughts", etc., as if they were suddenly required to clearly show subservience and obedience to the Neuron or making some kind of "opposite" statement as a protest for the audio record. D2B consumers know better than to even think criticisms of the supreme neuron leaders and the system of neuron D2BW secrecy.

How about the similarities to the big lies told under dictators like Adolf Hitler and Pol Pot? Nobody can talk publicly about remote neuron reading and writing as an actual scientific field, nobody can admit that light is made of particles that are material, and that light particles are the basis of all matter, the ridiculous non-Euclidean theory, and theory of relativity reign unquestioned with thousands of multi-million dollar "big lie theorists" constantly publishing page after page of text on the latest radical forces and particles of nature. The obvious fraud of the red-shift, as shown in the Humason 1936 image, shows not only the long-lived nature of the filthy lies under the neuron, but how even obvious truths can be easily suppressed from the public, and how easily all scientists and teachers can be made to echo the neuron party lie. Imagine yourself as a school teacher having to teach young people dishonest scientific theories that, as a D2B consumer, you know are not true. For a compassionate thinking person who cares about young people and education that must feel terrible. Beyond scientific lies, are the lies involving violent crimes, like the "19-hijackers brought down the 3 WTC buildings" and "Oswald shot JFK" lies, which are exactly parallel to the big lies of the worst dictators of history.

Brilliant excluded people telling the truth are remotely watched by millions of unseen people all day and night, producing billions of dollars for the supreme leader Neuron, but are dismissed as "kooks", "perverts", and "freaks", not receiving a penny of those billions they generate from the millions of one-way-only D2B voyeur viewers. Those "most watched" "kooks", "perverts", and "freaks" are never called what they actually are, which are "heroes", "popular democratic leaders", "very honest", "interesting", and "bright" humans.

There are many clear parallels between the current Neuron D2B system and the life-long monarchies of history. The Neuron is so powerful, that this comparison has never even been made publicly until now. Even this book is barely public and certainly will not be in wide distribution for a long time if ever.

Like any monstrous and murderous dictator, you don't have to look far to see all the victims of the Neuron. The thousands of WTC victims, the Kennedy's, Lennon, millions of galvanized victims- probably one of your friends or relatives was galvanized by old supreme leader "Kim Jong" Neuron, or will be.

Probably for any human born into great wealth and ownership of direct-to-brain remote neuron reading and writing technology, the natural tendency would be to stay with the same system of hording everything for themselves, and to not provide any D2BW to other humans; letting other people see and hear thought could only threaten their control over the public and over D2BW technology. But it seems likely that opening D2B to more people would increase the income of the system. Going public with D2BW would also add more variety to the people that can currently communicate through D2BW. Instead of a few people hogging seeing and hearing thought and neuron reading and writing technology all to themselves for centuries, isn't a better system letting everybody talk openly about D2BW, going public with it, and letting the majority control it?

Should a few individuals like AT&T's Randall "Kim Jong-il" Stephenson, who rule for life until their son Randall "Kim Jong-un" Stephenson takes over, be the people who get to decide which "Steve" gets to have "an in son", and which "Steve" lives a life of childless exclusion?

Is putting a singular, group of AT&T "Neuron" families, with life-long rule, that hordes and controls who gets to see and hear thought, and who gets to receive D2BW and who doesn't, the best system for a nation's and planet's communication service? Isn't a much fairer system a democratic system where the public decides who gets to see and hear thought, and who doesn't?

We can feel good about a few of us being able to know about, and talk somewhat openly about this terrible truth about the unseen Neuron pyramid scheme. But we have to accept that the current situation is terrible. You can see the potential future: there may be many more dismal centuries before the public finally even recognizes that RNRAW exists and has been secretly in use for centuries, then two or three hundred years more before talking about D2BW segregation is widespread and is a common discussion, and then a few more hundred years before most of the public gets even a very limited form of D2BW.

Are there any tips to getting included?

There isn't a lot I can say here other than: try to be in good physical shape, don't commit any crimes, and don't bite on suggestions that will make you look bad or do something unpopular. Try to fit in by understanding that the D2B consumers can see and hear your thoughts. Outside of that it's not clear what might help.

One question is: does talking openly about RNRAW and D2BW help, hurt, or have no effect on an excluded getting included? Some of your D2B "friends" will probably take a few thousand dollars to make a hand or finger over mouth gesture to you, "advising" you, as an honest and trusted buddy: "shut-up about D2B and you will get included just like me". But they get a lot of money to lie and mislead you, in particular if you trust them. Personally, I think we excluded are much better off talking about it, and trying to inform the excluded public. The D2B owners and consumers probably all know that 99% of those who are included stop talking about RNRAW once they are included. Probably most are older and simply try to pick up the pieces of their ruined lives, try to find a mate and make a family, etc. They probably quickly realize that an individual means nothing and is powerless against the massive all-powerful neuron industry. One theory is that, if you say nothing publicly about RNRAW and D2BW then you are "not a problem" (a NAP) for the Neuron, so they will probably ignore you and leave you as excluded for the rest of your life. But if you do talk a lot about it, then you are a problem for the Neuron, and they may feel compelled to try to silence you. Of course that could take the form of violent crime, and definitely does take the form of intimidation and threats of violence, but probably more likely it could take the form of trying to buy your silence with some kind of agreement. It may be that you fit into one of the two sides, as either a conservative or liberal, and so each side simply "includes" like-minded people at some time which they determine by majority vote of D2B consumers that receive their service. As an excluded who knows about D2B you probably will find that your D2B "friends" constantly say some word that is supposed to indicate when you will be included. But, knowing how many people are excluded, about Thane Cesar, 9/11, the expanding universe fraud- it seems more likely that those who control D2B mostly earn the nickname "inhumans". It seems likely that, despite all the D2B consumer promises of "we're going to get you", etc., that most people are excluded for life, because notice that the D2B *does not negotiate* with individual excluded people, for example, provide you with some limited D2BW service, offers of money, chat with you through D2B, etc. in exchange for you to stop talking about D2BW. My view is that they need to show me some regular D2B service before I will even think about offering anything in return. If you don't talk about D2B, and are excluded for the rest of your life, which is likely, then you did nothing to help end the segregation not only for yourself but for future generations of victims.

One possible truth may be that getting direct-to-brain windows may depend in some part on what "air-space" you live in, because it may be that the air around you is filled, of course, with nanometer size particle devices, but the majority of those devices, and nano-device space supremacy, may belong to the prevailing political party. So for example, if you are a liberal living in a conservative place, your chances of remaining excluded are probably higher, and the same may be true if you are a conservative living in a liberal place.

The initiation to D2BW process; when the D2B "gets" (includes) somebody

Clearly many humans get D2BW. So at some point in their life they must have the service turned on. What is the process for a person who is newly initiated to D2BW? Perhaps there are just more and more stronger and longer lasting D2B windows over time- perhaps starting with those during and waking up from sleep. I doubt that a video window just pops open directly in their eyes, and some guy starts talking directly into their ears. Perhaps, for many people, a D2B consumer, maybe a family member, sits down with them and goes over some well-worn script used to include people, like "you've reached an age where you are rcady to handle more responsibility…". Maybe there is an analogy for why the newly included person must keep everything about D2BW completely secret, from those excluded, but even from other D2B consumers. For example, they may use the example of homeless people; that there will always be people who are less fortunate than others, to help create reasons for the brutal secrecy. They must be very clear about the need for secrecy. Perhaps there is a very scary story, maybe about violence done to those who did not keep the secret. Then perhaps there is a slow introduction, like a thought-audio message from the

D2B consumer sent to the ear of the newly included like: "...do you hear my voice?...". Probably many times the newly included person says "yes" out loud, so the next line must be (again in thought-audio) "ok but the first rule to remember is that we can't answer out loud. Can you answer my question just in thought?". Then they probably introduce D2B windows. Probably there is a thought-audio statement like "you are about to receive a picture, directly to your eyes for a few seconds. It may seem uncomfortable at first, and you may feel that the picture is too close to you. You may feel uneasy because the picture moves with your head and you cannot turn it off. ...". Of course, I can only guess, but there must be some well worn introductory scripts. Similar scripts may be used initially with the public by a D2B device manufacturer or by government employees once D2BW goes public. There is a funny analogy with the first motion pictures shown to the public in theaters- one movie was of a train- and the public who had never seen motion pictures always leaned out of the way when the train came toward the screen.

Is there perhaps a lack of D2B resources that prevents most people from being able to get D2B?

The answer to this seems most likely to be "no". By the year 2000, like the Internet, the D2B technology must basically permeate the planet Earth and beyond. Seven billion people, the current human population of Earth, is not perhaps as large as many people might think. If each person was a byte of memory, they would only total 7 gigabytes; which is small by today's standards. But, certainly, there must be a limit on how much data can be stored, and for how long. The movie "The Final Cut" hints that only the most important movies of most average people's lives are permanently stored after their death- many are probably deleted to make space for those still living.

But in terms of direct-to-brain windows, there is no reason to deny the public this basic service. We all probably have RFID neuron organelles in our neurons and are "D2B-ready", but the D2B service is simply not "turned on" for us. Instead, D2B is used to murder, assault, molest and mislead millions of people.

Will there ever be a "big pay back" or "big turnaround" where the murderers fall and their victims rise to the top?
This is a question that I'm sure many excluded who realize that they are being denied seeing, hearing, and communicating by thought think about. Will there ever be a big payback? I don't know. Without a doubt, the possibility of excluded people themselves doing much is very low, because of the extreme and absolute advantage RNRAW gives to those who own and control it. Any reversal could only occur at the top: by neuron owners, and/or possibly by D2B consumers. But probably we excluded at the bottom can contribute a tiny influence over long periods of time.

There are historical examples of similar situations. One example is that of slavery and women's rights; clearly slaves were freed and both the freed slaves and women eventually won the right to vote, but there was no fast reversal or pay back for the centuries of injustice. But other examples include the US, French and Russian revolutions where the wealthier subjects, to a large extent, overthrew the old rulers and became the new rulers. So it's tough to really know. If I had to guess, I would say that it's going to be a slow and nonviolent progress into the future where more and more excluded get included and more D2B consumers start talking publicly about D2B. Of course, for many excluded the dream may be for a fast reversal: the D2B and 9/11 killers jailed, all the lies and liars exposed, all the victims compensated, and those who told the truth richly rewarded. There are cases where the sides have quickly reversed, like WW2, etc. Clearly it would be a case, in the USA, and probably all over the Earth, of the murderers (of the Kennedys, 9/11 victims, etc.) and liars being on the losing side, with all the victims of murder, assault, and remote molesting and their allies being on the winning and ruling side – controlling remote neuron reading and writing, showing all the thought-images and sounds of the past to the public, exposing and jailing the many murderers that are currently being protected by the neuron lie and secrecy, turning on D2B for the many excluded, and freeing the many wrongly imprisoned.

One wonderful aspect of any kind of reversal or turnaround is that the public will quickly see and know, for sure, who is a killer, who fought against the killers, who lied, and who told the truth. Even as an excluded, this can be seen simply by recognizing those who mislead the public about 9/11, and all the other big obvious lies. Perhaps the newly informed public might seek to reward all the new heroes: those who fought against the violent, who told the truth about 9/11, remote neuron reading and writing, and science, and who turned down the money to remotely murder, assault, molest, and to lie and shill – and seek to punish all the newly revealed scoundrels: those who did 9/11, that remotely murdered, assaulted, molested, etc. and that actively accessorized to try and trick the excluded from knowing the truth of many murders, and lies. It would be the end of many big lies and tricks because tricking and lying to people would be very difficult when all thought-images and thought-sounds are freely and widely shared.

One terrible problem is "forgiveness", as harsh as that may sound, but because forgiving is a central theme of Christianity, there has been a long tradition of forgiving those who murder. In my experience, forgiving and "turning the other cheek" is uniformly interpreted by the evil other side strictly as weakness and as an opportunity for more murders. The smart machines that the particle murderers use recognize this reality. We owe it to the many millions of victims to make sure that their murderers are punished and all the videos of their crimes made available to the public.

There may be an age, sometime in the near future, of many "big take-downs", where hundreds of people with many counts of remote particle murder, assault, and molestation are finally identified, removed from the particle technology, and jailed.

Each particle crime probably has an electronic record. Each murder, assault, muscle move, particle molestation, or money-window shill, are all transactions that must pass through the telecom companies' networks. Presumably each of those transactions is recorded and stored. If true, which seems likely, then that data, if it survives the transition from secrecy to being made public, can be used to punish those who have not yet been punished for their particle crimes, and for those people who committed crimes to be forced to pay back their victims. Such records must clearly show who are the particle murderers, assaulters and molesters, who paid them, how much was paid, in addition to who were the victims. Then clearly, all the criminals should be fairly punished, and all money involved given to each particular victim. I think most people can agree on punishing violent crimes, and compensating the victims with all money that was involved. In addition, all money gained for nonviolent remote molestation, and even for "shilling" may be voted to be paid to each particular excluded victim.

There are a lot of possibilities of a future where a majority of people try to "right" the "wrongs" done to D2B victims. One example of this is that the majority may vote to end the D2B read and write service for: those who voted "no" to going public with D2BW, for those with first strike violent crimes, and for those with multiple remote assaults or molestations.

Other reasonable thoughts and ideas about a future reversal include:

1) A fine for each proven remote molestation that equals all the money paid and at least $100 per second of molestation which must be paid from the molesters directly to the victim, in addition to the molestation being replayed on all involved in the funding, ordering and sending of the remote molestation.

2) At least 1% of all money paid to the neuron owners to see videos of a person or their thoughts should be paid directly to the person being watched.

3) Democratize at least one remote neuron read and write service for the public. Private people can make a competing neuron reading and writing service, but at least one should be majority rule, so that people who are Democrats, Green party, Libertarians, etc., are not excluded from using remote neuron reading and writing, simply because Republicans own the only neuron companies.

Basic wants for the future

Some basic wants for the future are to:

1) Stop the remote molesting (muscle moves, itches, unwanted thought-sounds, etc.). Let people see "who wrote it" to their brain, remove the neuron write permission from those who remotely molest, fine them and give the fine to the victim.

2) Go public with D2B technology, and demonstrate it publicly. Then everybody may not receive D2B but at least they can see that it exists, what it looks like, and how it works with many video examples.

3) Remove FSV (first strike violent) people from access to remote particle devices.

4) "get out"- educated D2B owners and consumers should focus on including more people that are currently excluded- in

particular the non-religious, evolutionists, anti-violent, pro-racial and gender integration, pro-consensual sex, educated, pro-democracy, those who have no problem with free-flow of all information and loss of privacy and secrecy, etc.)

Perhaps there could be a large law suit against the telecom companies, in particular AT&T. They clearly have broken many laws in their use of particle technology. Obviously the most serious are the violent crimes: **homicide**, **assault**, **conspiracy to commit homicide**, **accessory to homicide**, and then the non-violent crimes: **molestation** (done remotely on motor and sensory neurons with particle devices, which is still covered under the molestation laws and so is illegal presumably). Clearly the telecoms are violating the **privacy laws** in using micro and/or nano technology to capture and sell images and sounds from inside people's houses. Many nations have laws that clearly prevent other people, in particular non-governmental people, from capturing and selling images and sounds of people in their own homes. But there is no law, that I am aware of that forces people to compensate the subjects for images and sounds captured and sold of them. Another clear set of laws being broken by the telecoms are the **copyright laws**. Presumably all works made by any person are copyrighted, and the telecoms freely copy and distribute any and all works of people to D2B consumers, for money, without getting the permission of, or compensating, the owners of all those copyrighted works.

But beyond the clearly illegal activities, there is the massive discrimination being done against the "D2B excluded", which may not be illegal, and which has obscure but very real and enormous consequences that include inability to reproduce, discrimination in hiring, and the other abuses that come from many people viewing one person without that person having the opportunity to view those people watching them. Most likely any kind of law suit would be best done as a massive and collective effort by as many people as possible to try and make a permanent change of making D2BW available to the public, as available as cell phone service, and to compensate as many remote particle victims as possible. Beyond a law suit, perhaps ballot measures can be passed by the public, or new laws enacted by government representatives, to open and start to democratize direct-to-brain windows, and to identify and compensate the many victims.

When do you think D2BW and RNRAW will go public?

It seems likely that within the next century, and perhaps even within this or the next decade, remotely hearing what the ear hears and even hearing thought-audio will be made public. One of the authors of the 2008 paper which published the first public image of remote neuron reading (what the eyes see)[125], Yuki Kamitani, told me in an email message in 2009 that they already can distinguish between different syllables of thought-audio with magnetic resonance imaging, so in some sense hearing some aspect of thought-audio is already public knowledge. But also, the first public thought-image may be published within the next 20 years, probably by people in a University in Japan, or at the University of California at Berkeley. But from there it will be many years until it reaches the public in the form of handheld cameras that capture images and sounds of thought, and perhaps by then, D2BW will be available to the public.

One possible rule is that if one or more excluded people have figured something out, and are making some product or project, this is a signal to D2B owners that the technology or project the excluded

[125] Miyawaki, Y., Uchida, H., Yamashita, O., Sato, M., Morito, Y., Tanabe, H. C., Sadato, N., Kamitani, Y. (2008). "Visual image reconstruction from human brain activity using a combination of multi-scale local image decoders. ", Neuron, 60, 5, 915-929.
http://www.cell.com/neuron/abstract/S0896-6273(08)00958-6

are developing is "ripe" enough to allow wealthy D2B consumers to go public with the same technology or project on a larger, more well-funded scale. This may serve several purposes, for one thing, it takes the steam out of the excluded product and project- otherwise, the excluded might have a monopoly on the "new" idea or technology and have a large influence over the public, and also, the D2B owners can then step in and control the growth and distribution of the "new" technology or project. Some good examples are: walking robots (both with motors and artificial muscles), devices that can record sounds ears hear and thought audio, devices that capture images of what eyes see and thought images, intracellular wireless devices, microscopic flying wireless cameras, ships into orbit, and movies that tell the story of evolution, science and the future. Excluded people are slowly starting to use the available public technology to build walking robots, to hear thought, to build ships that can get into orbit, to make movies about evolution, etc. So, it seems very likely that the D2B owners may allow D2B consumers to step in, and capitalize or "cash in" on these kinds of projects before any excluded people might take the initial public thrill and excitement from the deprived excluded public when finally the "new" products and truths do reach them. For this reason, many D2B owners and consumers may write on talented excluded to try and force the technology and truths out to the public to gain D2B approval to go public on a larger scale. Either way, we all benefit- although the raw and hard-earned truths and technologies the poor excluded, as "worker bees", "reinvent" are left to history as relics of a dark age, they stimulate and force similar and much larger scale improvements to reach the larger excluded public. But wouldn't a much better system be for the D2B owners to lead the way into the future, instead of waiting for excluded people to "reinvent the wheel", before they take even a tiny step forward?

Knowing about remote neuron reading and writing is a huge advantage

As a person who now knows something about remote neuron reading and writing, you have a tremendous advantage over those poor people who know nothing about it. Just knowing about remote neuron reading and writing probably increases your chances of survival, of reproducing, of getting physical pleasure, of getting employed, of finding housing, etc. As a person who now knows about D2BW and RNRAW, but is probably a D2B excluded, you may now be viewed by many people as an "excluded who knows", which may be a step above the many other people who are marked as simply "excluded" (but still lower than a "D2B consumer"). The main advantage is that you probably will not bite on the many bad neuron written suggestions. But still, your chances for a date, a job, and reproduction are extremely low.

Once an excluded person knows about RNRAW, getting them to "bite" (repeat/act out) bad remotely written suggestions is much more difficult- it must be a similar phenomenon to remotely writing onto the brains of D2B consumers- because they know to reject violent and other inappropriate suggestions. But even so, there can still be "remote control humans". Does anybody who understands RNRAW doubt for a minute that, like our mouth or finger muscles are routinely moved for a second, that our entire bodies could not be remotely controlled indefinitely down to the last neuron? This truth creates a constant "Invasion of the Body Snatchers" phenomenon and fear in people, because like that classic horror movie, nobody can ever really be sure that their friend, mate, or even they will not be suddenly remotely controlled by unseen criminals of the opposite side, to involuntarily do violence or other unpleasant actions. I know from experience that to have a positive moment with a friend, instantly prompts violent and inappropriate images and suggestions onto our minds, always requiring an instant "fire back" in thought at the enemy, as odd as that sounds to those not familiar with RNRAW. When I speak or play music publicly, I have to constantly fire back in my mind at remote molesters-much of my life, and no doubt the lives of many other people, has been an endless "fire-back in thought at remote molesters" movie.

Just to give you an idea of how shockingly terrible the time we live in is, and in particular how far we have to go in terms of teaching the public about remote neuron reading and writing, I can say that, there was no phrase "remote neuron reading and writing" until I started using it on my web page, certainly by February 2010. Nor was the phrase "direct to brain windows", which I started using certainly by April 2011, ever stated out loud or published, to my knowledge, anywhere, publicly before. If you search the Internet for "remote neuron reading and writing" and "direct-to-brain windows" you will probably only find my web pages; even now in 2012, there is no public acknowledgement of the importance and secret of remote neuron reading and writing, or the massive secret society of those who receive direct-to-brain windows. How did I figure it out? It's possible that it was remotely neuron written onto my brain. I probably have received helpful remotely written hints (in between the endless remotely written garbage). As I said, I've never seen the phrase "remote neuron reading" or "writing" in print or heard it out loud ever in my life, and I have scoured the history of science, looking in particular at many physiological papers dealing with neurons. My best answer is that I haven't been handicapped by any religion, and also, in particular, that I have spent the last 8 years researching the history of science for my "ULSF" project. So this gave me an unusually good perspective on the course of science history as experienced and recorded by many of the great scientists of the past.

So, even now, to the best of my knowledge, and no doubt, for many years to come, there does not exist anywhere in print anything about remote neuron reading or writing or direct-to-brain windows. In fact, I don't know for sure what those who own and receive direct-to-brain windows even call it. I've called remote neuron reading devices "brain imaging machines". Another acronym I made is "SHASIASTAFB" (Seeing, hearing, and sending images and sounds to and from brains). Traditionally RNRAW has been called "telepathy", and now remote neuron reading is being done with "functional magnetic resonance imaging" (fMRI). It can also be called "remote neuron activation", and "direct-to-brain videos", or perhaps some other thing altogether. The phrase "direct-to-eyes" may sound less intrusive and crazy. An interesting analogy between the human brain and another electronic memory – a computer memory- that arises from the realization about remote neuron reading and writing, is the secret field of future technology called "remote transistor reading and writing" where even a tiny semiconductor transistor in a computer or robot somewhere can be remotely read from and written to using tiny nanometer flying or floating particle device networks.

Why do so few excluded people realize that they are excluded from D2B?

This is really unbelievable. The only answer I can give is that 1) they have never heard about remote neuron reading and writing- they know nothing about the scientific history and all the hinting, 2) they don't want other people to think that they are crazy by saying that they think people see them in their house, or hear their thoughts, and 3) they are remotely neuron written on to dismiss even the hint of such talk as unrealistic and/or unimportant. One constant unfulfilled dream of mine is to find an excluded woman that actually realizes the truth about direct-to-brain windows to talk to about it openly as a friend and possible mate to make a family with, and that a D2B included woman may step over the barrier to talk to me openly about direct-to-brain windows, again, even just as a friend, and, of course, also possibly as a mate. Sadly, this hasn't happen yet, but I'm hopeful and optimistic.

How could a secret like remote neuron reading and writing be kept secret for so long?

One answer is because the remote neuron phenomenon is a mixture of the evil, who murder and lie (9/11 being a fine example), the scared, who are too afraid to tell the excluded public the truth, the stupid, who support the evil murderers, and the under-informed excluded who can't recognize that the evil are murdering and lying. The collapse of science and the rise of the opposite side around 400 AD may explain the prolonged dishonesty, secrecy and idiocy. Well entrenched traditions of dishonesty and fraud have existed for centuries- religions being the main source of many supernatural and dishonest claims (Moses parting the red sea, Jesus turning water into wine, the many claimed "relics", psychics, horoscopes, fortune cookies, etc.). Suppression of science, honesty and education is a large part of the possibly 700 year delay. Making a muscle contract using electricity dates back (in public documents) to the 1600s, and remotely contracting a muscle to the 1700s, but yet the vast majority of people have never heard about this, and modern scientists have only recently (1959) even gone public with a pacemaker that can remotely fire a neuron. Another example of the lack of public education is that: in 1208 the theory that all matter is made of light was made public, but even today, this theory is not only not publicly accepted, but is not even publicly mentioned as a possibility, for example in the Encyclopedia Britannica. Wars have not helped to bring science to the public, in particular the most recent World Wars, during which most scientific publishing to the public stopped.

Beyond that, there is a strong feeling for many people to "do nothing", and also the feeling of "I get

D2B, so who cares about those that don't?". One reason why D2B consumers keep the secret is clear: because they don't want to lose the extremely valuable ability to see and hear thoughts. In addition, many people prefer an uneven playing field, where they have an advantage for mates and jobs, etc.- a newly included person might get a job or mate that you want. The more people that are included, the more competition there is for finite resources.

Imagine an alternative existence where we were born on a planet where everybody receives D2BW and talks about it openly, where everybody knows that all matter is made of material light particles and that our future is to build a star cluster, where everybody has already learned about evolution, where there is no antipleasure ferver, where enlightened full and constant democracy rules, where people routinely travel to other planets and other stars, where walking robots do all the manual labor, where violence and lies never happen: then we wouldn't have to spend endless hours writing books like this trying to tell people the most simple truths only to be looked at like we are crazy.

Shut up or speak up?

For myself, the answer is clearly "speak up". An analogy is: "Teach the public how to read and write, or keep the knowledge of how to read and write to yourself?"

Chapter 6
Secret Unpunished Violent Crimes

I am adding this chapter just to make sure that excluded people are aware of these massive and monstrous violent crimes that are not talked about publicly. Excluded people have been extremely under informed and we need to turn this around. The evil side has to trick people, and our side has to untrick them by showing them the truth. In addition, many of these crimes could only be committed and remain unacknowledged and unpunished with the help of the remote neuron reading and writing secret.

9/11/2001 was three controlled demolitions

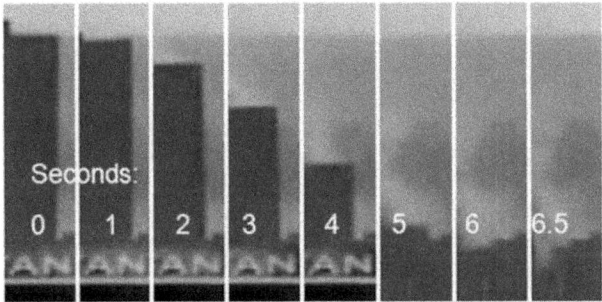

Figure 6.1. WTC7 falls into tiny fragments: all the steel joints broken symmetrically, in 6 seconds.

If you do not know already, the fact is that the mass murder of almost 3000 humans on 9/11/2001 was almost certainly controlled demolition. Simply put, no steel frame building, like WTC 1, 2 and 7, falls in 7 seconds into dust from an airplane collision, and WTC7 fell quickly without even being hit by a plane (fig. 6.1). In addition, there is clear evidence of molten metal that would be impossible from gas fires, and evidence of nanometer scale thermitic explosive orange and gray chips found in the dust of 9/11[126]. Photographs of the initial hole in the Pentagon show a hole that is far too small to fit a

---

[126] Harrit et al, "Active Thermitic Material Discovered in Dust from the

757 airplane through. Clearly Bush and Cheney (and no doubt many other people) planned 9/11 and then used it to justify the invasion of Afghanistan and Iraq. See the free Internet movies "In Plane Sight"[127], "Loose Change"[128], "9/11: Explosive Evidence"[129], "Zero: An Investigation into 9/11"[130], "9/11 Mysteries"[131], and the books and videos of David Ray Griffin[132], for more proof than any average thinking person could ever need. Remember that all the D2B consumers already know this.

9/11 World Trade Center Catastrophe", The Open Chemical Physics Journal, V2, 2009, p7-31.
http://www.benthamscience.com/open/tocpj/articles/V002/7TOCPJ.htm
[127] "In Plane Sight", Power Hour Productions, 2004
http://video.google.com/videoplay?docid=2361717427531377078
[128] "Loose Change", 2nd Edition, Korey Rowe, 2006.
http://www.youtube.com/watch?v=quTifldhH-g
[129] "9/11: Explosive Evidence", AE911Truth, 2011.
http://www.youtube.com/watch?v=Ddz2mw2vaEg
[130] "Zero: An Investigation into 9/11", Telemaco, 2008.
http://www.youtube.com/watch?v=8XRMrMdn0NQ
[131] "9/11 Mysteries", In the Wake Productions, 2006.
http://www.youtube.com/watch?v=2O7LwySqtr4
[132] David Ray Griffin 9/11 Lectures Playlist
http://www.youtube.com/playlist?list=PL8AAF0AB82519419E
David Ray Griffin books
http://www.amazon.com/David-Ray-Griffin/e/B000APTCK4

Frank Fiorini probably murdered JFK, not Lee Harvey Oswald

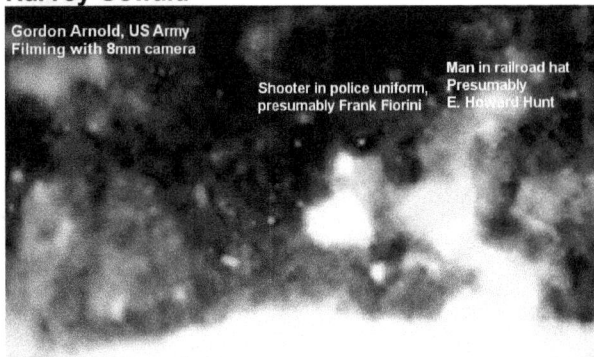

Figure 6.2. A magnified area of the Mary Moorman photograph- one photo says it all- but how about the thought and eye images too? The evidence is very clear- a man in a police uniform shot and killed JFK from the front.

Even most D2B excluded know this by now, but because it is so far in the past many young excluded people may not be aware. Fiorini, later caught burglarizing the Watergate with E. Howard Hunt and others, was never recognized or punished for this murder. Clearly the fatal shots came from the front, and were from the guy in the black police uniform next to the two other people (Gordon Arnold and probably E. Howard Hunt) in the Mary Moormon photo (fig. 6.2). Many of the honest D2B consumers use words that start with two "Γ"s to tell excluded that the initials of the killer are "F.F." which fits well with "Frank Fiorini"- a person who was funded and supervised by people such as George Bush Sr. and other people connected to the CIA.

Thane Eugene Cesar murdered RFK not Sirhan Sirhan

Figure 6.3.Thane Eugene Cesar (left) and Ted Charach (right) holding a photo of Thomas Noguchi showing the distance and position of the gun that caused the powder burns on RFK's body.

As an excluded, I, of course, initially believed the official story that Sirhan Sirhan killed RFK, but knowing the obvious truth about the JFK cover-up, I decided to spend just 15 minutes looking into the RFK murder. Within a few minutes I quickly found out what happened. This case is so simple and obvious. The Los Angeles County coroner Thomas Noguchi stated clearly that the murder weapon was no more than 3 *inches* from the lower right *back* side of the head of RFK (fig. 6.3), and all the witnesses say that Sirhan's gun was no less than 3 *feet* in *front* of RFK. Sirhan Sirhan definitely shot and wounded several people, but it's obvious that Thane Eugene Cesar, the openly anti-Kennedy and racist "Ace" security guard who was also working for Lockheed at the time, was the only person in the position needed to cause that kind of execution-style wound.

I interviewed Ted Charach[133,134], a journalist who co-produced the Golden Globe nominated 1973 movie "The Second Gun"[135], a movie that is the most

---

[133] Ted Huntington interview of Ted Charach, 07/12/2003. http://www.youtube.com/watch?v=HUhd_N52Po8
[134] Ted Huntington interview of Ted Charach, 01/17/2009. http://www.youtube.com/watch?v=3eAGJ6lAp_w
[135] "The Second Gun", American Films, Ltd, 1973.

accurate telling of who killed RFK that I know of, and which is being suppressed by the major movie distributors. See the links below for those free video interviews and "The Second Gun" for more details.

Neuron "Crime Creation" Department

It seems clear that there is a "Crime Creation" department of the Neuron: the Norway shooting, the recent Colorado "Batman movie" shooting, any and all violent crimes are certainly not only allowed to happen, but are clearly made to happen using remote neuron reading and writing. When you can see and hear thought and write to neurons, and have smart computer programs that can quickly identify humans planning violence simply by the images of violence on their thought-screen, how can you possibly not know about long term plans of violence? Then most obviously, how could the D2B owners and consumers not know who did any violent crime even long after the crime- as is the case for all "unsolved" murders? Remote neuron writing could be used to stop violence, but it appears that the exact opposite is the current reality: violent suggestions are routinely sent to excluded people, not the other way around- for example, countering naturally occurring violent thoughts with remote neuron writing. Perhaps this is due in some part to the popularity of violent videos among D2B consumers who pay lots of money to see them. But that is no excuse. We need total free info to see who is doing all the violence, including in the neuron, and then to actively stop them.

Chapter 7
Other Important Ideas

Just like excluded people have not been told mind-numbingly simple science truths like that all matter is made of light particles, so they have not been told mind-numbingly simple and logical social ideas.

First Strike Violence (FSV) is the big evil but somehow the public hasn't realized it

Violence is the big evil on Earth. Nonviolent activity (theft, trespassing, perjury, molestation, etc.) are a far lesser evil. But yet, the many murders of history and the apathy of the public to expose and try to stop them (in particular the most obvious ones like the JFK, RFK, and 9/11 murders) is evidence of this public's misplaced priorities. Unlike the victims of nonviolent crimes, the victims of FSV many times feel pain, suffer painful and lasting injuries, physical scars, and death. Many victims of violence are gone forever and aren't coming back, unlike victims of nonviolent crimes such as prostitution, recreational drug use, molestation, theft, etc. Nonviolent, although very annoying activities, like molestation, theft, lying, trespassing, copyright violation, etc. are not as serious as FSV activities. But much of the public has been hypnotized, or are slaves to tradition, in completely ignoring the most serious problem of first strike violence. For the current population of people, knowing who has sex-related crimes is a higher priority than knowing who has violent crimes. There is no "registry of violent offenders", and so as a result, people might hire, rent to, or date a person with multiple murders or assaults and have no idea, in particular if they are being excluded from D2BW. If forced to choose between spending a night locked in a room with a nonviolent offender or a violent offender, I think most logical people would choose the nonviolent offender.

Nonviolent activity that only hurts the self (drug and alcohol abuse, overeating) is a far lesser evil than FSV, but yet violent crime is of less concern than drug, pleasure and information related nonviolent "crimes". There are bizarre witch-trial-like sentences being handed out for those convicted with nonviolent crimes (drug use, pornography, molestation, copyright, violating national security, etc. offenses) even though the victims of the crime felt no pain and are living, while the victims of violent crimes felt a lot of pain and many are dead. Perhaps because they are dead and not around to seek justice, people tend to forget the crime. Perhaps nonviolent people are afraid to speak out against a violent person, because they fear becoming a victim of the violent person. Probably with the rise of science, logic, freedom of information, technology, and full and constant democracy where the public votes on all government decisions, the public will eventually realize that first strike violence is the biggest evil, and finally democratize, and create a logical punishment and sentencing structure, for the many various common crimes.

Perhaps many men may think that speaking out against violence is not manly, but stopping violence is the manly, courageous, and decent thing to do-it's staying silent that is cowardly and weak. We live in a time where fist fighting is obsolete, and people of any size or age can murder or seriously wound another person with the push of a button, and even remotely just with a "click" in thought. It's bizarre that hand guns are sold on the open market, but recreation drugs are illegal. Guns are used to cause far more damage, and D2B excluded people with a gun are a favorite target of D2B owners and consumers. The John Lennon, Giffords Safeway, Utøya Norway, and recent Colorado theater shootings and murders are all most likely classic examples of a D2B excluded hand gun owner being remotely controlled by remote neuron writing "suggestions" to do violence. Violence and murder done "indirectly" with an excluded, as opposed to directly with a particle device, and in particular, collectively spread among a large group of D2B conspirators and funders, creates more confusion about who to blame for a violent crime. The funders and executors of the neuron writing violent suggestions just cry "free speech" and say "we only showed them a picture of the crime...we didn't pull the trigger- they did!".

Just to make clear, I think that violence in self-defense from a violent attack is acceptable, of course, as most people can agree with. The long term desire though is to never have to resort to violence, by being able to stop violence before it occurs, which nanotechnology, remote neuron writing, robots, free information, and rapid full democracy is starting to enable. There are also complex examples where probably most people, myself included, would agree that violence is probably justified even though there is not an act of violence occurring, for example, somebody is pointing a loaded gun at another person, or where a person doing first-strike violence is imminent (within seconds). Many people believe that using violence (including murder) against a person who has broken into your house or apartment is justified. We should not forget that we all have loaded particle devices aimed at us all the time, so this fight against first strike violence is complex. Another example is firing back at those people who remotely molest people with particle devices; the molestation is non-violent, but it is annoying (making you think of a finger, contracting your muscles, making you itch, or to feel like you are getting an enema, etc.). When people write to your neurons, you can't write to your own neurons, and while it is nonviolent, it is very unpleasant and a violation of your body. The hope is for the remote molester's particle writing service to be stopped somehow, for a similar nonviolent but annoying molestation to be done back to them, and for all money involved, in addition to a fine ($100/second) to be transferred to the victim. There may be complex violence where many people vote and contribute money through their thought-screens to do violence, and a machine does the violence; with D2BW computers and nanotechnology, the list of conspirators and accessories of a violent crime can number in the thousands, but we need to expose, stop, and punish all of them.

I don't doubt that a "history of violence" movie has existed for a long time, although secretly, for D2B consumers- a movie that runs through all the major murders and shows clearly "who killed who". Such a movie should be assembled for the excluded public to reveal the truth about the many wrongly solved murders (certainly at least the most well-known of the wrongly solved murders, like those of JFK, RFK, and 9/11).

There is a simple truth: the more we expose, stop, and punish people who have done first strike violence (in particular those who remotely murder with particles, "galvanize"), the lower the chance there is of us being the victim of violence; the less we stop them, the higher the chance there is of us being the victim of violence.

Teach the public the details of evolution, the history of science, of the future, and of religions

I think that one reason people gravitate toward religions and creationism is simply because they have never been told and shown the history of evolution, science, and our possible future as owners and developers of a cluster of stars.

Most people know nothing or very little about the many details of evolution, for example that: 1) vertebrates (and insects) evolved internal fertilization (sexual intercourse) because on land (as opposed to in water) fertilizing the eggs directly had a selective advantage, 2) that one of our chordate worm-like ancestors evolved "upside down"- invertebrates like crustaceans and insects have their nerve chord near their front and not their back as we and the other chordates do, 3) that almost all fruits, nuts, and grains come from angiosperms, flowering plants (includes flowering trees), and that many fruit plants are closely related to each other- like "Fabales" (ℱ☉ℬⱯℒℰⱫ) the bean plants- most of the beans that we know are all closely related to each other; "Solanales" (SOLℬℕⱯℒℰⱫ), the green pepper, tomato and potato are all relatives, and "Cucurbitales" (ℋ𐐏ⱯKℝℬlⱵⱯℒℰⱫ)- cucumbers and squash are family. But flowers, which provide a large part of the food humans eat, only evolved around 150 million years ago, long after the first plant (1300 my {million years before now}), fungi (1200 my), animal (660 my), fish (560 my), insects (416 my), amphibians (375 my), reptiles (317 my), dinosaurs (228 my), and mammals (225 my). Birds (150 my) evolved around the same time as flowering plants. The list is very large, of very basic and interesting facts about the evolution of life on Earth that the public has not been shown and told on a large scale yet, like in a major motion picture called "Evolution", or as a mini-series like "Roots" on national television.

The public has purposely been not shown and told about the history of science. Most people cannot name many major scientists of the past (such as Aristarchus, Eratosthenes, Galvani, Franklin, Volta, Descartes, Newton, Faraday, Fraunhofer, Michelson, Mendeleev, Edison, Leavitt, etc.), or those who led (and lead) the struggle for human rights, but can name many people in sports, acting and music. In particular, I think many people would benefit from learning that everything is made of light particles, and hearing the details of remote neuron reading and writing, and about the development of microscopic and nanometer scale cameras, transmitters, and flying devices. It's stupid and brutal to leave the public so terribly under informed and misled. Beyond that, many of the theories being called popular science are actually deliberate lies (like the expanding universe theory, that D2BW doesn't exist yet, that people haven't figured out how to see, hear, and write sounds and images of thought yet, etc.). One myth being spread is that science is boring, but is that the truth, or just the voice in our head telling us that? In my opinion it's religion and most modern movies and television that are dull. I think the public would find very cute and funny recreations of the stories, for example, of Ctesibius playing the first organ, Vivaldi playing violin, William Hershel and his sister making lenses together for their telescopes, Trevithick driving the first steam carriage through town, the controversy of evolution, the women astronomers of Harvard; and find sad and tragic stories like the destruction of the Library of Alexandria (see the movie "Agora" for a recent re-enactment), the punishment of Galileo, the murder of Lavoisier, the burning of Joe Priestley's house, and smashing of the first spinning Jenny; and inspiring and emotionally stirring, how poor people like Michael Faraday, Thomas Edison, Marie Curie, George Washington Carver, Elizabeth Blackwell, and many others used their genius to rise up and succeed in life through science; but also the very fascinating truths and inventions figured out, for example the first rockets, paper, and printing in China, Galvani and the electrified moving frog legs, how crowds gathered to see the first proof that the Earth rotates, shown by the change in direction of the Earth under a pendulum set in motion by Michael Foucault, the first x-ray photos of Röntgen, the first public telephone of Reiss, and the first motorized flight by the two Wright brothers against all the critics who said it was impossible. Here again, a major motion picture called "Science", or a mini-series on television would help tremendously.

Perhaps part of the story of science is the story of our future, which many insiders must know but are deliberately and callously keeping secret from the public. Probably first in importance in my mind is that our future is to build a globular cluster, and that there is clearly a pattern of galaxy formation, from nebula, to spiral, to globular galaxy. Of course, I think many people would be amazed and pleasantly surprised 1) to see what cities on the Moon and Mars will look like, 2) how the Earth will be a beehive of swarming ships, 3) that the atmosphere of Jupiter will probably be consumed by many ships to reveal a massive molten hot surface, 4) that walking robots will be doing all the labor for humans, 5) that massive scale atomic transmutation of common atoms into more useful atoms will probably be used to provide our descendants with water and fuel, 6) to know what the moment that the first ships that reach the second closest Sun to us, Alpha Centauri, might be like, 7) to see that humans might actually reach the center of Earth and the other planets in the far future. Those are a few things I just thought of, there are many other interesting details about the future being denied. I can't believe that most excluded people would not be riveted by a major motion picture called "Future" that tells at least one of the many possible versions of this story, or perhaps a television mini-series on national television could enlighten a large majority of the public to these formerly secret great truths.

In addition to the history of evolution, science, and the future, much of the history of violence done in the name of religions is kept secret from the public, perhaps because the wealthy people who own much of the major media are religious, or fear that telling the truth may cause them to anger and lose income from religious people, but see books like "Holy Horrors" by James Haught[136] to get just a quick and minor introduction to some of the shocking and idiotic unprovoked violence done in the name of religions that much of the public never hear about. Here again, I think people on Earth would benefit significantly by seeing a history of the rise of religions and the brutality, conformity, violence, and destruction that has been done because of the many mistaken and radically inaccurate beliefs in religions. In particular the remote manipulation and tricking of many poor D2BW excluded people that have

---

[136] Books by James Haught
http://www.amazon.com/James-A.-Haught/e/B000APE82Q

common mistaken religious beliefs, through remote neuron writing should be shown.

Full and constant democracy where those who live under the laws get to vote directly on them

One of the most important and most simple of the "social" ideas that the D2B excluded are being denied, is the idea of a full and constant democracy where the people get to vote directly on the laws they have to live under. This idea is extremely simple and fair. There are a wide variety of implementations and specifics, but generally it is majority rule: the laws with the largest vote and largest majority having higher importance than those with less votes. One idea is that the local majority may have a higher value locally than the global or multistellar majority, so that people with overly conservative or overly progressive views can individually shape their smaller planets, moons, and cities without everybody having identical laws and values.

Many people I tell this idea to dismiss it for a variety of reasons, two of the most common being: 1) the public isn't qualified to understand law, 2) they don't trust electronic voting. But a fully democratic system where the public votes directly over the i-net on all government decisions seems inevitable to me. You don't need to have any special law degree to understand the issues involved in voting "yes" or "no" to "do we send our kids to invade Iraq?", and other similar recent government decisions. In terms of trusting electronic voting, I have to point to the credit card system as a clear example of how electronic transactions are routine and not corrupt. We trust electronic transactions to buy and sell products with credit cards, but can't trust a similar system to record votes? We could even be voting on important decisions (like to invade a nation) by allowing people to record their vote on paper, but knowing that people have been reading and writing thought for 700 years, I think we can use the paperless system. It may be that some kind of system already exists, but only for those who receive D2BW. Perhaps those few D2B excluded who know their vote might be recorded vote in their thought audio or on their thought-screen. It may be that wealthy people "buy votes" in the D2B, and it is viewed as the free market. Would you vote for somebody if they paid you $1000 to? Many people probably would.

Full democracy is the next logical progressive step up after monarchy and representative democracy. In a monarchy, the public have no say over the laws. The next step up from there is a representative democracy, where the public can vote for representatives who create and vote on the laws. The final step is the transition to a full and constant democracy where the public create and vote directly on all the laws they are subjected to. Imagine how many people that lied and cheated their way into the Presidency would have been stopped early on, and how many bad laws and murderous decisions could have been overruled by the public had full democracy occurred many years earlier. The less people that can see and vote, the easier it is to corrupt, the more people that can see and vote, the harder it is to get away with a lie or unpopular decision.

We need to let the public vote directly on all laws to make sure that those laws do, in fact, still have popular support. It's unfair to subject people to a law that they don't get to vote on, or that doesn't have the support of at least a 51% majority of those who must obey the law. Maybe some of those laws held popular support at one time, but don't anymore. Similar to the complaint of the colonists in America who said that they were the victim of "legislation without representation", creating laws that the public is subject to but that they don't get to vote on is "legislation without participation".

Many people don't realize many of the interesting consequences and details of full and constant democracy, because we've never been told about such an obvious system as "full democracy" by our major media. There are many advantages. First the public can vote out bad government employees, can overrule unpopular Supreme Court decisions and any President's unpopular decision (like a pardon). There would be a big, although progressive upheaval. The majority would remove the many existing unpopular laws, combing out the current jungle of laws into a nicely trimmed garden, and new laws would rise that are the most popular and universally agreed on.   Even taxes, budgets, salaries, health care systems, etc. could be determined by letting the public vote.

Currently, we excluded can only imagine how much the D2B owners and secrecy influence and control who and what the representative governments do. This is one reason we need to open and democratize at least one major neuron and phone service for the public. There can be non-democratic neuron communication companies, but there should be at least one democratic choice so that the undemocratic telecoms are not the only people who maintain a system of electronic voting, and to decide who gets to see and hear thoughts.

Another interesting aspect of full democracy is that the military would be democratized. For example, generals would not be appointed, but would be elected. All the employments and salaries in the military could be democratically voted on.

Just like the laws, even the court system will probably be fully democratized. The jury will simply be those who record a vote, court decisions are made instantly, and solidified or overthrown as more and more people vote. Probably "military" courts will be replaced with one democratic court system and set of laws for all people of each nation and planet.

In addition to the laws, military, and courts, the other departments like police and fire departments can be democratically voted on so the public can

decide who the best people for those departments
are.

Another strong argument for shifting from a
representative form of government to a democratic
form of government is that the public, when allowed
to vote on government decisions, will probably
budget their precious tax money much better than
representatives do. For example, look at what we
get for the billions of dollars in income, sales,
property, and other taxes: Free military "protection",
police, fire-fighting, school, roads, food stamps for
the very poor, social security after age 65, prison...
and that's about it- no free food, free drinks, free
health care, free clothes, free phone service (or free
D2BW service- or one that pays us 1% of all profits
made off of us), free transportation, free housing. I
think many people can agree, that the public would
probably promptly use those many billions in taxes
to make their lives more comfortable just like the
already ultra-wealthy representatives do. A trust-
worthy D2B consumer hinted that the votes of many,
and perhaps even all, US representatives are
routinely bought (probably through D2BW so there is
no trace)- a typical example might be those
Democrats who voted for an Iraq invasion under
Bush jr. knowing that 9/11 was three controlled
demolitions. In a free market even the votes of the
public can be bought by wealthy people and
organizations, so of course, even in a full
democracy, wealthy people would always have a
large influence on government decisions. Here's
another one of a million examples of how we don't
get our money's worth from a representative
democracy: I made "ULSF" in 8 years in my spare
time with no budget, but the United States
Government National Science Foundation, whose
job it is to promote science in the USA, and who
currently have a yearly budget of $7 *billion* dollars,
have not produced, to my knowledge, a single movie
telling the basic story of evolution, of the history of
science, or of the future of life of Earth. Like so much

of "representative" government, where the public doesn't get to vote on the decisions, our money does not provide us with any free services. I mentioned above that our tax money does provide us with "free military protection", but much of this money is apparently being used for murders like those of 9/11, for galvanizations of innocent people, and for the constant remote particle assault and molestation that many tax payers feel all the time. Most of our money goes into lying to us, and keeping everything a secret from us. We and earlier generations of citizens paid for much of the microtechnology of D2B, but we are not even allowed to see the technology we bought, access the movie libraries, or even use it to communicate with each other- and then, for all that money we and many other hard working people paid- not only do we not get to use it, but we are constantly *abused* by it!

Some people argue that full and constant democracy is the equivalent to anarchy, but that is inaccurate, because there is still a government with a full democracy. The only difference is that with full democracy, government decisions are made by a few million people instead of by a few hundred people.

Full and constant democracy is the end product of that natural feeling that replaced monarchy with representative democracy: majority rule, the most good for the most people.

Ending forced labor

One obvious simple truth the public is not being told is that we should view military employment as a job, not as slave-labor or indentured servitude. The first change is to allow people in the military to quit without any punishment. We can't imagine people being jailed or fined for quitting employment from the police, or from a restaurant, why should working for the government be any different? I can foresee other future improvements to the military the public might vote up including ending hazing (such as shouting at people training), ending the requirements of exercise and uniforms.

The other area of forced labor that should be voted down is in prisons. If people want to work in prison voluntarily that's fine, but my own vote is that they should not be forced to work.

Ending discrimination based on age

Since age is not a clear guide for all people, laws based on age should be voted down in my opinion. Many times a group of people lose their rights, because they are not powerful enough to defend them, and this is clearly the case for young people. There is a mistaken belief that young people cannot decide for themselves, but yet, I think there is a lot of evidence against this belief. For example, even a baby can express a distaste of like for some particular food. I think it is easy to see when a young person likes or dislikes something, for example a food, a song, a movie, a hug, etc. Young people should not be denied the right to make decisions that concern their own bodies, to vote, and to consensually work- this was the big problem with the laws that allowed slavery and that denied women their right to vote, own property, to get an education, to work, etc. The view that humans reach adulthood around age 16-18 doesn't align with the physical reality of when young people reach the adult stage (puberty) which actually occurs around age 10-12. But just because a human is young or old (or excluded for that matter) does not permit the denial of their basic human rights and the requirement of consent, or excuse the use of molestation or violence against them. For example, I am shocked that, while spanking of a nonviolent child is clearly a violation of the assault law, spanking and belting of nonviolent children is very common- in fact I was even spanked as a child. I would ask your parents if they were spanked- I was surprised to find that as a child my Mom was terribly belted by her Dad, simply for telling a dinner guest "don't pick at our turkey". Even in many developed nations it is an outrage and illegal to show a young human a picture of the nude human anatomy, but acceptable to assault them with a belt or by spanking.

Total freedom of all information-no jail or fine for any info owned- the myth of "privacy"

Here's another simple social idea that seems inevitable as we move into the future, but the major newspapers, television shows, etc. nobody, will talk about this obvious "elephant in the room", "emperor wears no clothes" kind of truth: the idea of a society of absolute and total freedom of all information and the consequences of that kind of system.

Clearly with the flying nano-cameras and neuron readers and writers the ancient idea of "privacy" is totally a myth, except maybe for the privacy of neuron owners from the excluded. With the advent of remote control microscopic cameras, microphones and particle devices, which may have occurred as early as the 1300s, the concept of privacy became obsolete (especially for poor people). The wealthy already see inside our houses, so stopping the free flow of information can only limit the poor, and those who don't get to see. The myth of privacy, only serves to protect the monopoly those wealthy people who routinely see inside houses and heads have held for many centuries and to protect those who murdered, assaulted, molested, stole and/or lied. The myth in many excluded people's minds that there is "privacy" is critical in order to get excluded people to do violence, theft, and inappropriate sex acts. If the excluded people knew the truth, that many millions see them, then, like so many D2B consumers, they never would follow remote suggestions to do violence, theft, or sexually inappropriate activity. Privacy is exactly the same as secrecy, and secrecy is wrong. Many violent crimes have only been seen by a small minority of people because of the widespread belief and support for secrecy and privacy. The much safer and smarter view, in terms of survival, I think, is the "leave no stone unturned" view- the view that nothing should be secret, and that secrecy is dishonest and evil. There is an obvious truth if owning or making public images of a crime is illegal because it violates a privacy, national security, pornography, or obscenity law, and that is that we cannot possibly stop a crime or catch a person who did a crime that we don't see, and that can only help those who did the crimes to stay unseen and unpunished to commit more crimes. Certainly when a person has done a violent crime, any "privacy" relating to images revealing the violent crime should be lost, and to keep such images secret is to be an accessory to violent crime. Opening up the free flow of information, in particular videos, is to close the space between the excluded public and the D2B owners and consumers.

One important reason to support total freedom of all information is because information is the only way to determine what the truth is. It's frightening how the major media and D2B consumers all lie about D2BW, Frank Fiorini, Thane Cesar, 9/11, 7/7, etc. You can see a potential even more frightening future where the D2B owners use their tremendous advantage to do more 9/11 kind of violence and massive lies, without ever been seen. For example, imagine the 9/11 crime without the videos of the three controlled demolitions; it would be much harder to prove controlled demolition. The same is true for the JFK murder without the Zapruder and Mary Moorman images, and for the RFK murder without the Noguchi autopsy; it would be much harder to know the actual truth. As if the block on the public seeing all the camera and thought images is not bad enough, D2B owners and consumers may even produce fake photos, for example in the case of the JFK murder and cover-up; the autopsy photos of JFK and the famous "Life" magazine photos of Oswald were altered. Beyond the probability of D2B owners and consumers doing more murders and cover-ups like 9/11, there is the very real possibility that people telling the truth about D2B, 9/11, the JFK murder, etc. might be jailed or hospitalized with made up crimes or psychiatric disorders by using D2B addicts to fabricate evidence and lie. Many poor D2B addicts can be easily made to lie in court in exchange for money or more D2B "services" like voyeur D2B videos, and free sex with beautiful women. I see this everyday with the D2B consumers that "shill" for money every 5 minutes. Probably the easiest way to remove those telling the truth about D2B, 9/11 and other lies is simply to galvanize them (remotely murder them). Those remote particle crimes must be very difficult to stop and to solve, and so total freedom of all information is very important for this reason alone. In addition, people telling the truth about D2B, 9/11, etc. and popular people that support full democracy ("political enemies", etc.) could easily be nonviolently removed from society by wealthy D2B owners and consumers using their tremendous D2B advantage. Hospitalizing or jailing people for made up non-violent psychiatric "disorders" or crimes is probably the easiest method to remove people from society, because you don't need any physical evidence like a corpse, bruise, or video, you only need a few D2B addicts to lie in court. Because of the lack of information, many times, even simply accusing a "political enemy" of some made up crime is enough, because the excluded people have no way of seeing the actual truth. Probably the easiest way to nonviolently remove people telling the truth from society is for:

5) A **drug crime**: D2B consumers in police simply claim they found illegal recreational drugs on the political enemy.

4) **Plotting violent crime**: the honest/political enemy is accused of plotting to do violence. No video or thought-images are needed; probably even a few fabricated text emails or photos of planted explosive materials can

be produced in court. This is the classic Bush-era "terrorist" charge. Nelson Mandela was subjected to charges like this.

3) A **national security crime**: the honest/political enemy is accused of violating national security. A recent example of this is the arrest of "WikiLeaks" leaker Bradley Manning. How could the D2B not see any potential leaker when they have a massive system of nanocams and D2BW? Then since this is a nonviolent "crime", why can't people who leak just be simply "let go" or removed from classified areas without being jailed? Many times, the leak is legal and ethical because it reveals a crime or lie.

2) A **sex crime**: one or more (many times young) D2B consumers lie in court about being touched sexually. No physical evidence, like a video of the crime, is needed. Alternatively, child pornography can be (electronically) planted on a political enemy or produced in court. Even having some D2B consumers produce fabricated text emails or text chat is enough to convict a truth-telling enemy of *attempting* to do a sex crime, for example with a person under the age of 18. Even the accusation of a sex crime is enough to ruin an enemy's popularity and career. This kind of phenomenon is often called a "media assassination", because the person accused many times can never recover their reputation even when the claim is totally false.

1) A **psychiatric disorder**: one or more D2B consumers in police simply transport the political enemy to a hospital where they are hospitalized for life. Most people will not defend somebody accused with a psychiatric disorder.

Without the public having access to the images captured from the many tiny cameras on streets and in houses, and without the public having access to all the thought images and sounds captured, you can see just how easily D2B owners, consumers, and the major media can trick the excluded public with lies. But with all those images reaching the public, the opposite is true; tricking the public with lies and fabricated evidence is much more difficult.

Decriminalize recreational drugs

Simply put, we don't jail people for being overweight, for drinking too much alcohol, that smoke tobacco, that don't exercise, or for living unhealthy lifestyles, and we shouldn't jail people that choose to use recreational drugs.

If we do jail people for owning or using recreational drugs, let it only be for a few days, until they become sober and have another chance at sobriety- not for years as are the current punishments. People who sell ('traffic") drugs can even get a death penalty in some nations –it's absolutely absurd. Guns cause far more damage, and then to *other* people, and we don't jail people who sell guns.

There are many arguments for decriminalizing recreational drugs. This is not to say that I think recreational drug use is a good activity; getting addicted to recreational drugs, like smoking tobacco, and drinking too much alcohol, is definitely a bad idea. But violent crime is the big evil, not nonviolent crime, in particular where the so-called crime is people who are only hurting themselves. Billions of our tax dollars are spent on arresting, jailing, feeding, and clothing millions of self-sufficient, non-violent, otherwise lawful people. That money could be used to expose and stop violent crimes and theft.

Decriminalizing recreational drugs, takes the "caviar" aspect away from recreational drugs- the feeling that expensive products must be very desirable, and it takes the million dollar profits out of a violent and illegal market- just like ending the prohibition on alcohol did. It's idiocy to support a system where poor young nonviolent kids spend hundreds of dollars, and subject themselves to dangerous areas and people, in the search for illegal recreational drugs. The war on drugs should be a nonviolent, non-jailing war that uses facts, videos, and other information to get the truth (in particular about RNRAW and D2BW) to the public so they can make good decisions about what to put into their own body.

I'm glad to say that I do not use recreational drugs, although I did when I was younger, and my advice to people is not to get addicted to recreational drugs, tobacco, or alcohol. I admit that challenging a person's mind or getting a different perspective on the universe by using some kind of drug, like a psychedelic mushroom or marijuana, might have some beneficial effect, and there are probably some illegal drugs that might have some health benefits. Recreational drugs are many times used to cure boredom. But an addiction to tobacco, alcohol, and many drugs is like a terrible handicap and a ball and chain on a human that may last for many years, and be very difficult to stop. For example, I hate that I smoked tobacco for years- that was so stupid, and probably one reason I didn't get a lot of dates. I think a lot of D2B excluded progressive-minded young people (which I was) are secretly and unconsensually remotely neuron written on with "ads" to start using alcohol, tobacco, and drugs at a young age – because it's a form of exterminating and making extinct progressives by the many conservatives that control D2B, but also because it produces more income for the companies that make and sell those products. Instead of trying to get a kiss, the excluded young people end up sucking on a bottle or cigarette. Partying with alcohol and drugs is a terrible and stupid tradition, but like the traditions of the religions, which are clearly based on lies and are also unhealthy and stupid, millions of people still believe in them and follow them, no matter how unpleasant and self-hurting they are. A drunk person is an opportunity for the remote neuron writers to make a person, especially a young sexually frustrated excluded person, to do something stupid using remote suggestions, in particular violence and/or something sexually inappropriate. I think many people turn to alcohol, tobacco, and drugs not only because that is the wishes of the conservative remote neuron writers (and of those companies that sell alcohol, tobacco, and drugs), but because it's an alternative to physical pleasure, which in this time is callously forbidden. I think that people use alcohol and recreational drugs, not only because they are excluded and can't defend against remotely neuron written suggestions, but to try to escape from the terrible reality of our situation – where murderers and bullies run the show and the decent, smart, and honest are being trampled on by them – basically – this constant theme of the Inquisition seeking to torture and exterminate the young Galileos and Galileas of today. Many excluded people (progressives and non-religious) are remotely funneled into alcohol, tobacco and drug addictions and away from reproduction by the violent liars that control remote neuron writing. All most young people really want is physical pleasure- but because talking about and even educating people about pleasure, dating, kissing, etc. is not allowed, that natural desire is diverted into violence, alcohol, partying, sports, and other more acceptable non-pleasure/non-sexual based activities. In addition, pursuing sober intellectual pleasure and science is shunned by the religious conservative majority as being "nerdy", but yet, embracing and exploring science and technology seems like the logical path of the future, to wealth, friendships, and intellectual stimulation – really the exact opposite of religions (except for maybe friendships). So it's terrible that young people are not finding a large and solid tradition based not on religions, violence, sports, and partying with alcohol and drugs, but based on science, stopping, exposing, and punishing violence, showing and telling the stories of history, science, and the future, using their talents to express their views, celebrating physical pleasure, physical fitness, and other natural, sober-minded, and honest pursuits.

Decriminalize consensual adult prostitution

If we could see the direct-to-brain community, I am sure we would see a massive market of physical pleasure for money- perhaps the biggest money making market of D2B (or perhaps the buying of remote writes, or reads themselves gain the telecoms the most money). They apparently already have it down to a science with people getting "kiss", "hand", "sucks", "coatings", "holes", etc. I can only report from what I hear as an excluded. But beyond the fact that the D2B owners and consumers have already accepted a free and open market of pleasure for money, are all the moral arguments for not punishing people who do pleasure for money.

It's absurd to punish people who are involved with pleasure for money. First, these are nonviolent and consensual activities. One argument constantly echoed is that there is not consent. But agreeing to provide physical pleasure for money is exactly as consensual as agreeing to bag groceries, drive a truck, or fight in a boxing ring for money.

Beyond this, in the next few decades two-leg walking robots are going to be doing all manual labor (cleaning, shopping, driving, etc.). Humans doing physical work is going to become more and more obsolete. But even after many centuries, there will still be one job that humans are still hired to do – only one manual labor task will remain- not cooking, driving, or cleaning – but prostitution – providing physical pleasure. Not that the robots will not be very good at providing physical pleasure that looks and feels exactly the same if not better than a human, but certainly many humans will prefer to touch other actual humans for physical pleasure. So isn't that ironic that the only employment that will last long into the future is currently illegal?

I think that many educated people do not realize how harsh the society we live in currently is. For example, people (even children) can fight each other for free and for money (as sport), but even adults *asking* each other to pay or receive money for pleasure can result in being jailed. We live in a bizarre and shockingly frigid anti-pleasure group and tradition. We are all the product of sex- nobody can deny that – but yet many people are so violently antisexual. It makes no sense to curse the source of our life and physical pleasure. For some reason religions have always taken an extremely anti-pleasure based view. While violence is ok and manly, physical pleasure and even talking about physical pleasure is seen as a weakness. It's tough to know why this mistaken belief that physical pleasure is somehow wrong or a weakness occurred. Maybe population control was an important concern at the time. It may be that many people seek to stop others from enjoying pleasure out of jealousy. There is a popular view that sex and most forms of physical pleasure are extremely serious and life altering events that should occur secretly and only between two people that mate for life. But a more accurate view is that wanting physical pleasure is not a weakness but a natural and normal desire, that there is nothing wrong with the nude human body in public, and that most forms of physical pleasure (like sleeping together, fondling, kissing, or "handing") cannot result in pregnancy and so are not nearly as dramatically serious as many people make them out to be. So I think that in comparison to violence, people have come down overly harshly on no-chance-of-pregnancy nonviolent forms of consensual physical pleasure.

In the time we live in, people can be jailed even for asking to pay to touch a genital or to be masturbated ("solicitation"), or for even owning pictures of nude humans ("pornography"). For years people could be

jailed for "seduction" (having sex under the promise
of marriage). In many ultra-religious nations people
who are suspected of homosexuality are routinely
executed- it's bizarre. Violence is ok, but even
talking about gentle touching is a big evil in their
minds. But what we are seeing is a gradual
enlightenment taking place on the Earth: the old
laws that jailed people for homosexuality, adultery,
seduction, etc. are falling because the public is
becoming more logical and educated in its views of
sex.

Many of the people in prostitution are poor people,
and they could be using money from prostitution to
pay for food and for an apartment. So those who are
so vocal in opposition to decriminalizing prostitution,
perhaps without knowing it, are actually helping to
starve and make homeless many perfectly fine
people, who otherwise might have had a job and
would have survived.

Another point is that it seems illogical that making
pornography, where people are paid to have sex, is
legal, but prostitution is not legal. Clearly, there is
almost no difference between if an individual pays
for sex for themselves, or if a filmmaker pays for
other people to have sex.

People buy stories in the media all the time to try to
link the pleasure market with children, forced labor,
and violence- which may happen, just like any
market, but we don't hear about forced labor in other
markets. There is also never any mention of the
beneficial effects of legal prostitution: that otherwise
unemployed homeless people are working and able
to buy food and rent a room, that there is less
violence because aggressive males have less
sexual frustration, and that without the secrecy,
people can more openly monitor and stop crimes
and the spread of sexually transmitted diseases.

Just like advocating decriminalizing drugs, just
because a person advocates decriminalizing
prostitution, doesn't mean that they are involved in it.
In the time we live in, very few people openly

advocate decriminalizing adult prostitution, because the stigma of being labeled a "pervert", "whore", or "pedophile" is so great, and because they don't want to appear to be a person who is involved with prostitution even if they aren't involved in prostitution. In addition I think that many wealthy D2B owners and D2B consumers may prefer keeping adult prostitution illegal because it may lessen reproduction among poor and D2B excluded people. Beyond that, the telecoms ("Madam Bell") probably want to maintain their secret monopoly over prostitution under D2BW; ending the prohibition on prostitution would obviously result in a loss of profit for them, as both sellers and buyers would have an alternative system. There are very few heroes to point to who stand against the bizarre anti-prostitution fervor. Many nations have already stopped jailing people involved in consensual adult prostitution (Wikipedia has a map assembled[137]). The National Organization for Women passed a resolution calling for decriminalizing prostitution in 1973[138], and in 1999 the United Nations CEDAW committee called for decriminalizing prostitution in China.[139]

Antipleasure ferver

One mystery about sexuality is how the vast majority of people alive are the product of two people who had sex, but yet a strong majority view exists that views nudity, the nude human anatomy, and public images of sex as being highly offensive; generally sex and sexuality are viewed as a bad thing. The obvious irony and paradox is: How do people so openly outspoken against pleasure and sex exist? If they were truly anti-sexual, they would

---

[137] http://en.wikipedia.org/wiki/Prostitution

[138] Weitzer, R.J. Legalizing Prostitution: From Illicit Vice to Lawful Business. New York University Press, 2011.
http://books.google.com/books?id=cjJKlRQsEv0C&pg=PA249

[139] Pitcher, J., M. O'Neill, and T. Sanders. Prostitution: Sex Work, Policy and Politics. SAGE Publications, 2009.
http://books.google.com/books?id=_L-UkxjWVoAC&pg=PA101

not have sex, and not reproduce and therefore would quickly go extinct, but yet, here they are in large numbers. If they were so anti-sexual would they not be very frigid and cold and very difficult to cuddle up with, and therefore very difficult to have sex with, and wouldn't that always result in extinction of the anti-sexual person? So it's mysterious. It may even imply, that some aspect of why sex continues is deeply linked to dishonesty, trickery, and hypocrisy- that is, sexual success is linked to people who publicly support one view, but secretly practice the opposite view. This truth implies, for example, that to be a successful reproducer, you must pretend not only that you are not sexual, and not interested in sex, but that you are highly offended by all things sexual, like pornography, and must vigorously label others "perverts" in order to distance such labels from being attached to you. But then, in a passionate moment, suddenly, hypocritically you must throw away and betraying all your antisexual views to make passionate love to some member of the opposite gender (kind of like Majors Burns and Houlihan in "M*A*S*H"). With all due respect to religious people and D2B consumers, of whom there are many honest people held hostage to tradition, religions have been breeding true for "liars" for centuries and the D2B secret and lies have amplified the success of the dishonest.

One clear truth is that there is an extremely unnatural selective advantage for reproduction by those who control and receive direct-to-brain windows. Over the last few centuries, those who received D2B must have been able to reproduce far more easily than those who did not receive D2B. Like wealth, being a D2B consumer plays a large part in determining if a human will reproduce. The problem is that progressive, non-religious, and overly honest people are being systematically excluded, and so perhaps that is one reason this dishonest public antisexual fervor (but private sexual fervor) is so popular: because dishonesty is a requirement for receiving D2BW and receiving D2BW is linked to success in reproduction.

Just like nonviolent atheists, agnostics, and scientists were brutally punished in the Inquisition, and other witch trials, and Jewish people in Nazi Germany, so now in our time the "sex offender" is the new heretic and witch. Shockingly, there is no "violent offender" hysteria. There is a "sex offender" list, but no "violent offender" list. Sex offenders on this "black list" are not separated into "violent" and "nonviolent", and they are all put together on one list whether they have one or one hundred sex offenses.

Remote neuron writing is currently being used to a large extent with the goal of making sex offenders of excluded. Using remote neuron writing to make a person "bite" on an inappropriate sexual suggestion is one of the easiest ways to lower the value of a D2B excluded, and to put them on the path to extinction. Excluded are constantly bombarded with thousands of suggestions to inappropriately touch a person under the age of 18, or to go outside in the nude, etc. But the neuron writers are never viewed as sex offenders- all blame is placed on the excluded human that bit on the suggestion.

One focus of the antisexuals of this time is on child pornography. Making child pornography illegal only protects those who abuse children. I think that making child porno illegal is also popular among the wealthy neuron owners because they want a monopoly on all information, and they are nervous about the public seeing images of what their eyes have seen, and the images on their thought-screens which may show their involvement in violent particle crimes. One interesting aspect is how AT&T and the telecoms always mysteriously escape all blame for crimes committed on their wires and with their equipment. For example, the government and public never blame AT&T for illegally transmitting child porn. Are we to believe that AT&T doesn't know what is on their wires and wireless network? Even if they don't (which is very doubtful given RNRAW), are they not partially responsible as accessories to the child porn crimes?

When we excluded see the news of young kids being killed in sex crimes (Chelsea King, Adam Walsh, Megan Kanka, Jon Benet, etc.) should we not wonder, as crazy as it sounds, "how could a child be murdered when people have been seeing and hearing thoughts with dust-sized cameras and particle devices for centuries?"? It can only be that the owners of remote neuron reading and writing are allowing these murders to occur. It may be that the view of many neuron owners and consumers is that sacrificing one poor child is a small price to pay, for the greater good of using the anger such a murder creates to pass new laws restricting the freeflow of information, and which create new options to jail or intimidate their enemies with made-up, hard to prove or disprove, career ruining, sex crimes. Much of remote neuron violent crimes are about causing a "wave of indignation" among the excluded public. Imagine if the public ever does get to see D2BW how many people that were accused of made up sex crimes will be shown to be completely innocent, and how pissed they will be at all those people who actually caused all the crimes remotely using particle devices. One thing is clear, that the money for sex and violence-filled videos fuels AT&T and the neuron. D2B consumers must pay millions of dollars to the telecom companies to see videos of the murder of those kids, their eyes, their thought-screen, the thought-screen of the murderers, their thought-audio, the thought-audio of the murderers, different cameras that captured the murder from different angles, the sounds of the murder, the profiles or those who did the remote neuron writing, etc. – we can only guess- but I don't think we have to guess much.

Making child porno illegal is helping to protect many people, some of whom are probably D2B consumers and perhaps even D2B owners- because nobody in the public, in particular excluded people, can see their crimes.

With child pornography, like images of violent crimes, we need to remember that capturing or seeing evidence of a crime should never be considered a crime, only those who do a crime- in particular, a violent crime, should be punished. Protecting evidence of crimes, in particular violent crimes, is of extreme importance for people to know who really did the violent crimes. In addition, making sure that all the images can be easily seen by the public, and not just by employees of governments is very important. Thousands of important videos of crimes have been confiscated, kept secret, and destroyed by people in government, the Zapruder film, the Scott Enyart film, and the 9/11 videos, are just some famous examples.

It seems likely that many wealthy people want to fuel the public obsession with sex "crimes", in order to remove the more logical focus on violence. With the focus on sex crimes, the possibility becomes less and less of ever seeing images of the people that did the controlled demolitions of 9/11, and a million neuron murderers that currently live unseen and on the loose.

Trying to make child porn illegal is a disaster. For example, can you imagine that all the ancient vases showing child pornography might be illegal? We have the label "paedophile" but not "paedokrust" (a person who is violent to children), which shows the misplaced priorities. And what about child "krustography" or "violentography"? Is the next step to make viewing, buying, or selling images and evidence of violent crimes against humans under the age of 18 illegal?

Clearly humans start masturbating around the time of puberty; eleven or twelve years old. I was eleven years old was when I started masturbating – how old were you when you started masturbating? Masturbating is healthy and natural. What is unusual is if a person does *not* masturbate. For example, the practice of celibacy, common, in many religions, to me, seems biologically unnatural, unpleasant, and unnecessary. The vast majority of humans have masturbated, but yet most people do not publicly acknowledge that they have ever masturbated. Health professionals should be recommending that post pubescent people masturbate or have sex (with careful attention to pregnancy and disease) regularly at least once a week, (for example to masturbate or have sex every other day at some regular time), but they are mostly silent about masturbation, orgasm, and sex. Most people are not even told, for example, the theory that you should try to postpone your orgasm as long as possible in the last few minutes, to savor and have more control over the actual moment of orgasm. To deny young people the physical pleasure of kissing, and mutual masturbation is a form of neglect and child abuse because they are denied their right to consensual physical pleasure with similar-aged peers. Many people get a pet to substitute for this severe restriction, but it seems more natural to allow young people to kiss and fondle each other in a non-pregnant making way. The remote neuron has great pleasure in making young horny people do embarrassing things in desperation for physical pleasure- for example making them masturbate during class, or getting them to try to asphyxiate themselves during masturbation- which is a shocking and terrible thing that some young people do in lonely and isolated excluded desperation with unseen inhuman remote neuron writers constantly writing such suggestions for D2B consumers hungry for shocking videos. I am so sorry to say that many D2B owners and consumers must pay a lot to the telecoms to see videos of these poor young sexually frustrated excluded kids biting on terrible and radical remotely neuron written sexual suggestions. It's brutal, but you can't blame the D2B consumers, you need to blame those writing the suggestions and those enforcing total and absolute celibacy on minors- the very humans that probably need love and caressing the most.

It's scary even to talk about the anti-sexual fervor, because, like so many witch hunts of the past, any people who stand up against the injustice and idiocy to defend those labeled as witches are quickly labeled as being witches themselves, and publicly standing up for children's right to consensual physical pleasure with other similar-aged kids, and for protecting and making public images of crimes against children, is no exception. Those very few who are brave and disgusted enough to say anything are promptly and inaccurately labeled perverts and paedophiles, while the non-stop remote neuron, massive, billion dollar, secret, crime video-production, telecom industry continues production without missing a step.

I sometimes hear labels like "whore" or "slut", and it's evidence of a massive anti-women equality group, because sex is not just the responsibility of the woman, "it takes two to tango", but also, because a human should enjoy sex and orgasm once a day, and at least once a week- there is nothing wrong with having regular sex. A monogamous person may have just as much sex as a person (female or male) that has sex with a different partner each day, so there is no physical difference. But also with labels like "whore" and "slut", as a male, it's annoying, because, women don't need to be chastised and made colder in this ice age we live in- they need to be warmed up, turned on, complimented, celebrated (not celibated), encouraged to enjoy their bodies, and pursue pleasure soberly and intelligently. The "whore" people are usually the same people that constantly call everybody "gay", "fag" and "dyke". I'm so looking forward to the day when people realize that homosexuality is of no concern and is completely normal and natural. Violence is the big evil and should be the big taboo. Then, to those people, absolutely everybody is gay- it's so annoying- it's like homosexuality is their primary focus in life. I often say when hiring- "anybody but the violent, rude, and the anti-gay, for the love of the work environment." Defending gay people is just like defending people accused of being a heretic or witch- because you are promptly labeled gay by the idiotic antigay. In truth all people are probably naturally bisexual to a certain degree- there are only three kinds of people- those who admit to having masturbated to same gender touching, liars, or those who have had their minds chained with fear. Constantly labeling the excluded celibates "gay" and "pervert" has been a patented method used to stop the breeding of the honest for centuries.

Much of the anti-pornography, anti-sexual ferver is clearly anti-education, because how can you teach young people about anatomy and about the 600 million years of animal evolution through sexual reproduction when it is illegal to show young people images that contain a penis or a vagina? Do we want the next generation of doctors and other professionals knowing next to nothing about vertebrate anatomy?

There is a startling truth that most of the public is not told, and that is that we are descended from protists (single-celled organisms) that probably were very much like a sperm and ovum. Our sexual organs are the most primitive parts of our body. Before there was a brain, muscles, or intestine, there was a gonad. It's amazing to realize that all the other organs are, in some sense, "accessories". The entire digestive, nervous, muscular, skeletal, circulatory, respiratory, and endocrine systems are later developments; products of 600 million years of the elaborate mating dance, and evolved just because they are effective at bringing together sperm and ovum, those cells so like our primitive protist ancestor.

Consensual-Only Health Care (Ending torture and unconsensual experimentation in the psychiatric hospitals)

It seems clear that the psychiatric system will eventually have to end all unconsensual electrocuting, surgery, drugging, and bodily restraints. The psychiatric system is a scary phenomenon, in particular because a person can just be picked up off the street and held indefinitely without any trial or crime being committed. Many people call the psychiatric system a "Siberia" after the famous destination of many political prisoners under Stalin. The first murdering by poison gas done by the Nazis was in psychiatric hospitals on "patients" (many probably pro-democratic, homosexual, and Jewish people) as a "hygienic measure". Holding people who haven't committed any crime or have only committed a misdemeanor, in a building against their will for an indefinite period of time is a violation of the Habeas corpus act - and that was progressive in the 1200s. Injecting people with drugs (experimental or otherwise) without clear consent, and physically restraining people (in particular nonviolent people) with four point limb restraints, to a bed for extended periods of time, leaving them unable to move their four limbs freely, and with no choice but to urinate and defecate on themselves, violates the ancient constitutional amendment and basic human value that forbids "cruel and unusual punishment", in particular where the punishment far out-weighs whatever crime was committed.

Just like lowering the popularity of an excluded can be done, very easily, by neuron writing constant suggestions to do something sexually inappropriate, so can making excluded do unusual things that get them locked in a psych hospital, which dramatically lowers their popularity, and attaches the lifelong labels of "psycho", and "nutter", etc. to them- even when their "crazy" act was the result of remote neuron writing, was nonviolent, and only a misdemeanor (like a temporary public outburst), etc. Many of the people targeted for this kind of remote abuse are the honest and educated people of the D2B excluded. You can see how the 9/11 controlled demolition and Kennedy killers constantly try to label those telling the truth about those crimes as "crazy", "nuts", "kooks", etc.; trying to link the telling of truth with a psychiatric disorder. This is also the case for those that are figuring out remote neuron reading and writing, and starting to tell people publicly and honestly that they think people can hear their thoughts, that they hear voices in their head, that somebody is remotely moving their muscles, etc. You can see clearly how the abstract theories of psychology, coupled with the stigma of being labeled with a psychiatric disorder are being used to suppress the truth about remote neuron reading and writing, and to protect many murderers, like those who planned and carried out the controlled demolitions of 9/11, the Kennedy killings, and many other murders. They are people that have to use labels and theories of psychology, because the facts and physical evidence don't work in their favor.

Many of the claims based on psychology are very abstract and have little or no basis in physical science. Take for example, labels like "psychosis", "neurosis", and "manic depression". Unlike a broken bone, cancer, or virus like HIV, there is no bio-chemical diagnostic test to show that a person has psychosis, neurosis, or manic depression.

It's tough to state clearly what the basis of the "crazy" phenomenon and theory is. For example, one aspect is that a person who is labeled crazy has "inaccurate" beliefs. But we never hear the word "inaccurate" being used. The problem with locking up people with inaccurate beliefs is that, having inaccurate beliefs is not only completely legal, as it should be, but historically, a majority of people have always had extremely inaccurate beliefs- for years people insisted that the Sun goes around the Earth, the main claims of all the religions, that Moses parted the Red Sea, that Jesus turned water into wine and brought people back to life, are all obviously inaccurate- but we don't jail religious people, restrain and drug them, and try to deprogram them with the more accurate stories of evolution, science and our possible future as a globular cluster. I think that I've shown quite clearly that the claims of a big bang expanding universe with background "radiation" are inaccurate, and that light is material and not an "electromagnetic" wave, and so even many currently popular claims of science are inaccurate and "crazy"- but physical restraints, injections and other brutal punishments are not the answer- showing the public good information is a far less destructive and far more effective answer for correcting inaccurate beliefs.

Another aspect is that a crazy person is thought to have unusual and unpredictable behavior. Some people are somewhat unpredictable, or have inconsistent behavior, but as long as they are not violent, that shouldn't be viewed as being reason to be hospitalized. Much of psychology is geared towards removing creativity and difference in society. For example there is an attention deficit disorder, but having an attention surplus is not a disorder yet. The labels and "disorders" of psychology favor the dull; being overly dull is not a disorder. Beyond that, many people who do unusual, or unpleasant things are the victims of remote neuron writing, and would not do unusual and unpleasant things if there were no evil idiots writing terrible suggestions on their neurons.

In light of knowing the secret of remote neuron reading and writing, and that much of thought is simply thought-images and thought-sounds, most of the theories of the famous people in psychology (Freud, Jung, etc.) about "id" and "conscious", etc. have to been seen as far removed from being accurate. No psychology textbook talks about thought-screen, thought-audio, etc. – but remote neuron reading and writing has been known by many of those authors for over two hundred years, so clearly, without talking about remote neuron reading and writing, much of the theories of psychology are mostly fraud or too abstract and outdated to be of any value.

Many times the person labeled with a psychiatric disorder is a nonviolent person to begin with. Ultimately, in my mind, violence is the big evil, not non-violent activity, and those who do violence should be jailed, not hospitalized. There can be consensual treatments offered to violent people; for example, I think one good video to show them might be about how excluded people can be remotely neuron written on to do violence for some D2B consumer who doesn't receive any of the blame.

How about the so-called cures offered for the "crazy": drugs, physical restraint, etc. Everybody that is taken to a psychiatric hospital gets drugs; as if a drug is going to magically teach a person about the history of science, about how people write to their neurons, etc. Then think of the bizarre system we have: people who use recreational drugs consensually are brutally jailed for years, but those who just say "no" and refuse psychiatric drugs in the hospitals can be injected with drugs against their will. Almost every psych hospital routinely uses four point restraint torture without consent. This is simply a "punishment", not a treatment, and is clearly designed to make a person less aggressive and more submissive – many times because they have to beg "please untie me" and "please loosen these straps", and "please let me use the bathroom", etc.

It's shocking that those who murder can avoid prison by pleading "insanity", and then can be released as "cured" from some abstract and unprovable "psychiatric disease". The "insanity" defense is ridiculous in my opinion. My vote is for people to be held accountable for their crimes no matter what their motive was. In addition people who work in a hospital should not be subjected to violent people. We need to clearly separate prison from hospital and violent from non-violent. Beyond that, all health care should be consent only.

Many times, remote neuron writing is used to make people do inappropriate things (yell in public, talk back to a person in police, go out in the nude, do sexually inappropriate things, etc.). It will be interesting to see how many of the people locked in psychiatric hospitals are there because they are D2B excluded and bit on some remotely neuron written suggestion. Probably the vast majority of people locked in psychiatric hospitals are D2B excluded- people who know absolutely nothing about remote neuron reading and writing. So in terms of treatments, wouldn't that be the first thing to tell them about? "You know it may be that some terrible violent criminals are able to remotely write to your neurons…and it's best if you realize that any voice you hear or image you see in your thoughts is not from God but is from these really terrible criminals of the 9/11 demolition Kennedy killing remote galvanizing kind…". But that's not being done.

One kind of funny, but dangerous truth about the harsh, fraud-filled, time we live in, is that a person can never simply be "wrong" or "inaccurate" on some claim- instead, in the eyes of many people, having a mistaken view is always apparently symptomatic of a systematic psychiatric disease that results in them being "crazy". Only rarely is a person ever viewed as just being wrong on a few theories.

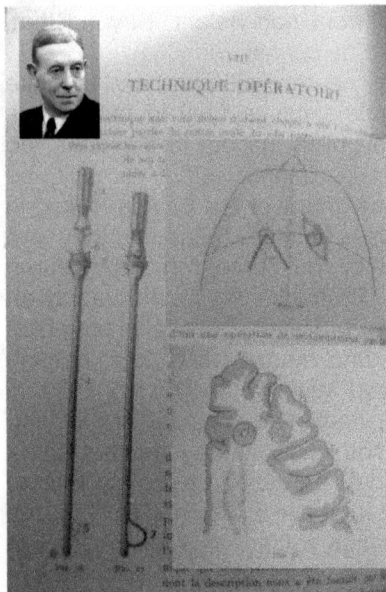

Figure 7.1. Nobel Prize winner Egaz Moniz and images from his paper detailing the first use of the Lobotomy.[140]

Beyond all this, the history of psychology is terrible. One bizarre story is how the person who first performed the "lobotomy" (perhaps unconsensually) on some people, Egaz Moniz (fig. 7.1), won a Nobel Prize for the "leucotomy" (lobotomy)[141] in 1949, at a time when the truth about how psychiatric hospitals were used to euthanize innocent people in World War II, and the newly created "Nuremberg Code" advocating consent-only experimentation were still fresh in many people's minds. The lobotomy is a procedure that simply crudely removes parts of the

[140] E. Moniz, "Tentatives opératoires dans le traitement de certaines psychoses" (Tentative methods in the treatment of certain psychoses), Paris : Masson, 1936.
also in:
J Am Med Assoc. 1937;108(21):1828.
http://jama.ama-assn.org/cgi/content/summary/108/21/1828-g
[141] "Egas Moniz - Biography". Nobelprize.org. 29 Oct 2010
http://nobelprize.org/nobel_prizes/medicine/laureates/1949/moniz-bio.html

brain. Even today, nonviolent, lawful, humans are subjected by law to being electrocuted with the radical and doubtful treatment of electroconvulsive therapy. For centuries people accused with abstract and unlikely "disorders" have been subjected to bizarre and brutal "treatments" like being restrained in a tub of water, having teeth pulled, etc. There are numerous books on this history, but no major psychology books or professionals want to educate the public about the shockingly brutal and violent past of psychology. Things are starting to change, however, as more victims are telling their stories, and more videos are reaching the public- for example there is now even a museum showing the history of psychiatric torture in Los Angeles. What we really need is a major motion picture or mini-series on national television that gently and inoffensively describes a lot of this truth and suppressed history to the excluded public.

Let the record reflect that in our time, labels of psychiatric mental disorders were used to suppress many great scientific and criminal truths, like the truth about the big lies of our time: remote neuron reading and writing, the "red shift" of the "expanding" universe, and the murders of JFK, RFK, 7/7 and 9/11. All those telling the truth are dismissed with abstract, meaningless, labels like "crackpot", "crank", "bonkers", "wacko", "nut", "kook", etc. Simply put, we live in a time where truth is called crazy and crazy is called truth. The murderers and their accessories have to draw on these abstract labels, because the facts and physical evidence don't support their inaccurate claims.

A "One-letter-equals-one-sound", one-stroke, easy-to-write, democratically-determined phonetic alphabet

We live in an era of massive language mispronunciation, corruption and misspellings, because we do not use a "one letter equals one sound only" alphabet. So as a result, there is a constant mystery, in particular, for young people just learning language, about how to pronounce any word. For example, should the letter "a" be pronounced like the "a" in "ape", or like the letter "a" in "apple"? Another result is that people are inaccurately pronouncing words like "Caesar" (Greek: "Καίσαρας") and "circle" (Greek: "κύκλος") because those words originally had a "k" sound, but were then changed to use the letter "c" which can have two sounds. When we make a spelling mistake many times people may look at us like we are stupid and uneducated, spelling correctly is many times viewed as a measure of intelligence, but isn't it more stupid to reject a one-letter equals one-sound alphabet that would, at once, end the vast majority of spelling mistakes, and provide more ease in spelling and reading for all people? So, a one-letter-equals-one-sound alphabet would solve all of these frustrating problems: the actual pronunciation of words can be more accurately stored over the years, kids can more easily learn language without wasting precious hours learning unique spellings, and questions of how to pronounce any word would be instantly removed. In addition, since all languages humans speak (Arabic, Chinese, English, French, German, Hindi, Italian, Japanese, Javanese, Korean, Russian, Spanish, Vietnamese, etc.) all basically use the same exact sounds (the "a" sound in "ape", the "B" sound in "ball", for example), a single alphabet can be used for all human languages. In fact, this truth is a piece of evidence that all the 30 or so basic sounds of all human languages originated in Africa before the common ancestors of all Homo sapiens now living moved out of Africa into Eurasia (perhaps 100,000 years ago), and then on to Australia and America, because human communities separated by large distances have different words, but not different base sounds. A single international easy to use alphabet could simplify communication between people of different languages. The stories of every language could more easily be understood by all the people of Earth, because the task of learning the sound of each word would be removed. There is an International phonetic alphabet, but those letters are not single letters (for example "eye" is represented with two letters "ai" instead of simply "I"), and not single stroke letters. The fact that sound frequency can change the meaning of words in Chinese adds complications, but perhaps people can figure out a simple and logical system for that, like adding one of five slanted lines before or over a word. Here is a sample one-letter-equals-one-sound alphabet I made a few years ago:

ABDΣFGh⊦JKLMNOPRSTUVW႘႗ZACႸ⊔႗JLMN⊖
RSⴀⴅⴑⴏ

I designed the letters to be writable with just one stroke. Just 39 sounds and their symbols cover most words of all human languages.

Some words with their native spelling, and with a one-letter-equals-one-sound phonetic alphabet spelling are listed here:

Language	Native alphabet	Phonetic alphabet
English	cat	KAT
English	teach	TEC
English	hello	hƎLO
French	bonjour	BONJUR
Chinese	你好	NE hⵔU
Russian	привет	PREVYƎT
Arabic	ابحرم	MⵔRhⵔBⵔ
Persian	ملاس	SⵔLⵔM
Hindi	नमस्ते	NⵔMƎSTA
Italian	ciao	CⵔU
Vietnamese	chào	CⵔU
Japanese	こんにちは	KONECEWⵔ
Korean	안녕하세요	ⵔNYⵔNhⵔSAO
Hebrew	םולש	SⵔLOM
Greek	γειά σου	YASU

It's tough to know how thought-images and sounds might replace text as D2BW becomes more widespread- an image can many times be a faster way to communicate information than with text- and certainly there is no need to create thousands of symbols to represent each of the infinite number of sounds possible. It seems likely that instant translation of all major languages can be written to the eyes of those who are not denied D2BW, and probably soon walking robot assistants will translate and communicate directly with people of different languages for their human owners.

The Other Species, their thoughts, vegetarian alternatives

Of the many monstrous results of keeping D2B secret, is that all the thought images and sounds of the other species have been kept from the public for centuries. Which species can form thought-images is not even publicly stated yet. When a D2B consumer walks by any of the other species, a bird, dog, cat, horse, etc., probably they see the little thought screen and hear thought audio too- like a dog remembering a song, or the eye image of a bird looking at you. Seeing the thought images and hearing the sounds of the other species might probably lower the popularity of eating them by humans.

It's terrible that most people have not even been told such simple truths as: "leather is the skin of a cow", and "meat is muscle". I myself didn't make those two connections for years.

I think that some tastes are adapted to, because I remember clearly how the first time I ate a hamburger, after the first bite, which tasted like hay, I asked my Dad "This is what everyone eats?" in disbelief. I have been happily vegetarian for over 10 years and now after a massive rise in the popularity of the vegetarian diet, almost every meat product has a vegetarian equivalent at a similar cost. I think that if there is not a big difference between the meat and veggie food item, then of course, it's better to have the veggie food item because no animal is killed or enslaved for it.

An actual logical non-pseudo-science way to lose weight

There are so many myths and pseudo sciences surrounding weight loss, and many remote suggestions of images and smells of food, that I thought I would tell excluded people the system I used to lose 100 pounds. It's extremely simple: I only eat two meals a day spaced about 12 hours apart (6am and 6pm) with absolutely no other snacks except gum and any liquids I want to drink. I eventually even stopped the gum chewing habit by recognizing and sometimes "firing back" in my mind at each (presumably) remotely written impulse/reminder. One added benefit is that I am hungry in the morning and that is a strong motivation to get up and out of bed even if I'm still tired.

There is a lot that motivates me to get into good shape: to look attractive, to have a longer life, to attract a female partner, to increase the chance of getting affection (kissing, dates- already a very low probability for an excluded, in particular in this bizarre anti-pleasure age), to enjoy sex, (such that it exists in this ice age era), and to make a family.

When I eat, I eat whatever I want- but initially I actually counted bites. It's a simple equation: your body can process maybe 100 bites of food a day (50/meal), so if you eat under 100 bites of food you will lose a little weight, if you eat over 100 bites you will gain a little weight. I stopped counting bites, but a camera and computer could make it more convenient and "bite count" is good info to track along with weighing your body every day to know if you are gaining or losing weight. The key is to take small bites, and to savor the food. Since there is so little food, it tastes better when eating. Sometimes I imagine eating whatever I want to and it's amazing how similar to the real experience it is. In addition, I find that this greatly reduces my budget for food, so that I can actually spend more money on more exotic and interesting food items. Then I do a very minimum exercise every day of simply jogging 1 mile and sit-ups. On alternate days I also exercise my chest and side muscles. The key is stretching the muscles where the fat is. For males it's almost exclusively the abdomen and side muscles – why oh why are we not told this simple truth? For females probably repeatedly using the buttocks muscles is the most efficient method to lose the most weight from the area with all the fat. It's pointless to focus on muscles that don't have fat built up if the goal is simply to lose fat. You know that a particular exercise is effective when the muscles where the fat is become sore.

Reality of bipedal robots doing all manual labor

I talked about this earlier, but it is something that people have to understand and plan for. The idea of a "job economy" where humans work is falling to the past. Currently, humans are born, go to school, get a job, make a family, and then retire. But in the future, probably people will not have to get a job. It seems certain that in the future, clearly, robots will do almost all manual labor (cleaning, planting, harvesting, packaging, driving, shopping, etc.). How and when is not clear, but perhaps humans will democratically create some minimum standard of living, for example: all humans get a room, one meal a day, a free set of clothes, etc. If a person wants more, then they will have to find some way of getting money. Perhaps pleasure for money will be one of the few jobs still open to humans, but the vast majority of manual labor tasks will be done by low cost bipedal (humanoid) robots. On the plus side, many of the repetitive mindless assembly line tasks now being done by humans will be done by machines. Jobs like that are much better suited for machines – it's cruel to subject humans to that kind of work by keeping robots a secret. Instead of having regular jobs, people may get money from the government just to live on, in addition to money from wealthy people for their votes or because they have similar views and want to promote those views, etc.

Chapter 8
Other Popular Mistaken Theories and Beliefs

Throughout history there have been many mistaken theories and beliefs, and my desire is not to put-down, or ridicule those who believe inaccurate or alternative theories, but to try and reach people with what I think are the more accurate theories. One classic theory was that Earth was the center of the universe, which of course, has fallen to the belief that Earth is only a small planet going around a Sun which is one of the many stars in the Milky Way Galaxy. The theory that all of space is filled with an "aether" fell out of favor, and I think the Big Bang Expanding Universe Theory will also eventually fall out of favor with the public- in particular when they see that many of these popular claims are all part of the "big neuron lie"- many people lying because they are held hostage by the D2BW owners. Once you realize that many of these explanations from those in the media, in science, in crime solving, in religions- in every field- are part of that big neuron lie, then you realize that the actual truth of many events and the universe in general may be very different from the popular explanations told to the excluded public.

Here are some other popular mistaken theories and beliefs:

That God or Gods exist

Believe in a God or Gods if you want to- all people must be free to think and believe what they want to freely. Without trying to upset or offend anybody, there are good arguments that no God or Gods actually exist:

- For centuries there was only polytheism (belief in many Gods), so you can see, historically, the origin of the "God" theory (the theory that a group of Gods control the universe), created by humans sometime less than 100,000 years ago. Monotheism, the theory that only one God exists also developed recently, around 3,350 years ago (1350 BC) starting with the Egyptian pharaoh Akhenaton.[142] Judeism is no more than 3000 years old, Christianity only 2000 years old, and Islam only 1400 years old.

- There have been so many Gods throughout history, and each religion has their own God or Gods, how could only one be the correct God? As one smart statement says- we are all atheists of some God; atheists have just rejected belief in one more God. I often say that belief in atheism is more conservative than Christianity and Islam because atheism is older than both those "new age" religions.
- We should interpret the universe in terms of matter and space. We should only believe in those things that can be observed with our senses.

This is not to say that there is nothing that does not defy logic. For example, that there is no beginning or end in time or space in the Universe, or that galaxies and living objects exist at all. And this is not to say that I do not have the deepest respect for the Universe, which we are a tiny part of, for all matter and space, and all that we have learned about the Universe. I feel a strong and natural need to be true to my beliefs and feelings, and am in awe of the Universe and all that science has shown us. What most people call "God" would probably most closely map to "the Universe" in my value system.

---

[142] "Akhenaton." The Columbia Electronic Encyclopedia, Sixth Edition. Columbia University Press., 2012. *Answers.com* 27 Oct. 2012. http://www.answers.com/topic/akhenaton

Isn't it ridiculous that so many people promptly pray whenever tragedy occurs? We can't wait around for a God or Gods that don't exist to solve our problems; we have to solve our problems ourselves. Here we are reaching an age where the public is going to be communicating to each other with thought-images and thought-sounds, but there are still so many people that believe in these extremely ancient and inaccurate theories and traditions.

That a Heaven or Hell exists

Many poor people are terrified by the thought of a Hell, but it is obvious to me that the myth of Hell is a recent invention, and that no Hell exists in the Universe. If you say that something is "old as Hell", you are saying that something is actually very young relatively speaking, because Hell as a concept is no older than 1000 B.C.E. We can't close our eyes and hope evil goes away; if we really want to stop evil, we have to be brave enough to examine it. I encourage you to research this history too; anybody can see the truth of this simply from recorded history. The myth of Hell was adapted from the earlier myth of an "Underworld" ruled by the God "Hades" (and referred to as "Hades"), which was located underground in the Earth. Hades is the ancestor of Hell. Unlike Hell, Hades was not just for bad people, but was simply (and inaccurately) thought to be where people go when they die.[143] Judaism, which Christianity was initially a form of, had a similar concept called "Sheol" where dead people were thought to go after death.[144]

I recently learned that the infamously scary and supposedly unlucky and evil number of the "beast", "666" was actually, before Christianity, thought, by

[143] "Hades", The Concise Oxford Companion to Classical Literature, Oxford University Press, 1993, 2003. *Answers.com* 16 Sep. 2012.
http://www.answers.com/topic/hades
[144] "Hades", *Encyclopædia Britannica, Encyclopædia Britannica Online.* Encyclopædia Britannica Inc., 2012. Web. 16 Sep. 2012
http://www.britannica.com/EBchecked/topic/251093/Hades

Greek people, at least, to be a "lucky" number, in particular a lucky dice roll of three sixes, a "triple six".[145] Isn't that a funny truth? Oh a 666! It's my lucky day!

Beyond that, rejecting religious theories is no big evil. First strike violence against non-violent people is the big evil on Earth; nonviolent activity can only be a lesser evil. But rejecting false claims, religious or otherwise, is no evil at all, and is a great good.

The myth of a Heaven has a similar story. For centuries before Christianity and Islam, Heaven was the home of the Gods; mere mortals like humans were not destined to go to Heaven after death (except for a very few "heroes").[146] Only later did this myth change to make Heaven a place where "good" people go, and Hell a place where "bad" people go after they die. How can so many people accept so primitive and inaccurate a theory as being "divine" or sent by God, in particular, knowing that humans recently changed the theory and "renovated" Heaven?

That a Devil exists

While there certainly are many bad humans living on Earth (first strike violent people, and those who constantly remotely molest and lie), there is no Devil, Demon, Angel, or God of evil that is the supervisor and controller of all evil as we humans define evil on Earth and In the Universe. Many people may not realize this, because the subject matter itself frightens many people, and many people that do examine it are wrongly and unfairly "demonized", but like the theory of a Hell, the theory of a Devil is a recent invention. The earliest mention of a Devil is of a "Satan" in the Hebrew Bible around 600 B.C.E., in

---

[145] Aeschylus, The Agamemnon, Choephori, and Eumenides of Aeschyles, tr. into Engl. Verse, 1865, p4.
http://books.google.com/books?id=CIYCAAAAQAAJ&pg=PA4
[146] "heaven", *Encyclopædia Britannica. Encyclopædia Britannica Online,* Encyclopædia Britannica Inc., 2012. Web. 16 Sep. 2012
http://www.britannica.com/EBchecked/topic/258844/heaven.

the book of Job, and initially the Satan of the Hebrew Bible works under God's supervision, and is not the ruler of all things evil[147,148].

That Jesus was the son of God, or a part of God, or was supernatural

- Jesus really made no significant contributions to life on Earth, and in particular science. People who lived before Jesus gave the world pottery, the wheel, writing, the correct size of the Earth, etc. and humans after Jesus did much more, inventing the electric light, the automobile, the airplane, the camera- we have benefitted much more from their hard work than from anything Jesus, Muhammad, Abraham, Siddhartha Gautama, or Confucius did.

- It is ridiculous to idolize and view as relevant to today the teachings of a person that lived over a thousand years ago, in a time when there was no electricity, running water, Internet, airplanes, thought-audio recordings, etc.

- It seems not very smart to view a human that lived 2000 years ago as being somehow radically different from the trillions of other humans that have ever lived- in particular when you think of the last 200,000 years of human reproduction and expansion.

- It's extremely Old Worldly to live a person's life devoted to a guy who lived thousands of years ago. Similarly, most people don't wear animal hides and use stone tools anymore.

- Many people have not really examined the history of Christianity. For example, if Jesus lived, he was definitely a Jewish person practicing Judaism. Few people recognize

[147] Pagels, "The Origin of Satan", 1995, p39.
[148] "job new 2". The Columbia Electronic Encyclopedia, Sixth Edition. Columbia University Press., 2003. Answers.com. http://www.answers.com/topic/job-new-2

that Christianity is a form or "sect" of Judaism. Christianity was initially Judaism and then made later adaptations.

- History is filled with the stories of violence done against innocent people in the name of Christianity. The Inquisition was used to inflict very violent punishments, like burning people alive. Most of the time the victims were nonviolent people, and often the brightest and most honest people of Earth (like Galileo and Giordano Bruno). In addition, the violence and discrimination against people just because they have Jewish ancestry is a common brutal and illogical theme throughout the entire history of the Christian religion. Any Christian anti-Jewish racism is ironic and illogical, in particular because the founder of Christianity was a Jewish man.

- Jesus may not have even existed- other people that lived during the same time left writings like Pliny the Elder, Livy, Paterculus, Strabo, and Josephus. If Jesus did live, he left no writings, and perhaps didn't even know how to read or write.

- Many of the claims of Christianity are obviously false, like that Jesus brought people back to life, that Jesus visited people after his death, etc.

- There are great humans who made tremendous contributions to life of Earth, but it seems foolish to spend large portions of a person's life focused on a single person. How better it seems to me, to focus on the big picture- of evolution, the history of science, what we need to do to go to the other stars, and what our future might be like, but then also to focus on our own personal life and pleasures. We shouldn't torture ourselves doing activities that are not fun and interesting. Isn't time better spent trying to find a good mate, a good meal, and to pursue things that really interest us and that give us the greatest pleasure?

That Muhammad was a profit of God

- As is the case for Jesus, Muhammad was just another one of the billions of humans.
- Almost all the same above arguments explaining why it doesn't seem smart to spend a person's life centered on a human who lived over a thousand years ago, apply in the case of Muhammad and Islam. Many other people have contributed much more to science and to making life easier for many humans than Muhammad did. It's absurd to worship and view a person's writings made in a time without electricity, airplanes, computers, etc., as relating to modern life.
- Having to bow to Mecca five times a day, like prayer, has no actual benefit, and is a tremendous waste of precious time.
- As is the case for Christianity, there is a lot of violence done under the name of Islam, and like Christianity, the violence done under Islam is often directed against the brightest minds, and against nonviolent people for trivial nonviolent "crimes" like blasphemy, adultery, etc.

That many claims of Judaism, Buddhism, Hinduism, and other religions are accurate and useful

- Many claims of Judaism are obviously false, like that the Universe was created in seven days, every part of the story of Adam and Eve, the claim that Moses parted the Red Sea, that people must follow specific rituals to please the God, that sending thought-audio messages to God (prayer) will help to solve problems, etc.

- In terms of Buddhism, many claims are simply superstitious and almost certainly false, for example, that putting gold leaf on a statue of Buddha will bring good fortune. Like a "lucky four-leaf clover", there is no truth to those claims, and it doesn't help to participate in and perpetuate those inaccurate beliefs.
- The theory of reincarnation, one claim in Hinduism, seems very doubtful to me. I also doubt the theory of Karma, although there may be some truth to the idea that a living object may receive collective benefits or losses as a result of their individual actions- if ever humans on Earth choose to suddenly embrace logic.

I encourage people to learn about the details of evolution (for example in the books "Prehistoric Life"[149] and "The Ancestor's Tale"[150]), the history of science (for example in "Asimov's Biographical Encyclopedia of Science and Technology"[151]), and basic history (like "Compact History of the World"[152]). It seems obvious to me that science is a much more honest, logical, accurate, and interesting philosophy to embrace, follow, and participate in, than religion is.

Superstitions
Superstitions are terrible, and all of them are simply false. For example some classics are: a black cat crossing your path is bad luck, breaking a mirror causes seven years of bad luck, a severed rabbit foot is lucky, certain minerals have healing or magical powers, Friday the 13th is unlucky, opening an umbrella inside will bring bad luck, crossing your

[149] Palmer, et al, "Prehistoric Life", DK Publishing, 2009.
[150] Dawkins, "The Ancestor's Tale", Houghton Mifflin Harcourt, 2004.
[151] Asimov, "Asimov's Biographical Encyclopedia of Science and Technology ...", Doubleday, 1982.
[152] Parker, "Compact History of the World", Barnes & Noble Books, 2002.

fingers will bring good luck. There is no logical basis for any of these superstitious claims, and we shouldn't perpetuate them.

Many other mistaken popular beliefs

There simply are a lot of people who create lies and a lot of people who believe them. That many humans are easily tricked into believing lies has created many, many mistaken theories and labels for nouns or objects that simply don't exist in the universe. Just to name a very few: Gods, Heaven, Devil, Hell, Witches, Soul, Goblins, Ghosts, Angels, Hades, Zombies, Santa, Fairies, Magic, Luck, etc. Many of these old beliefs were formed long ago in the past by people who had never flown in a plane, had no electricity, running water, or telephone, etc.; people with very primitive views compared to modern times. To a certain extent, some ideas were precursors to modern science beliefs- for example the ancient concept of "soul" was created before people modernized anatomy, identifying the heart, the brain, etc. But there are also many non-existent claimed objects that were the product of science, for example: an aether that fills all space, phlogiston, n-rays, and black holes. I don't believe in the existence of any object (noun) for which there is no physical, observable evidence. In the absence of any physical evidence for an object, I think the smartest view is to presume that it does not exist in the universe, and that some people long in the past mistakenly thought it did, or simply created a pretend creature or place.

I'm a somewhat slow person. For example, my "ULSF" project has taken 8 years to complete. But, 700 years of waiting for RNRAW and D2BW to be made public is absolutely a shocking, ridiculous, and torturously slow delay. The same is true for the fact that 2000 years later many people are still talking about Jesus as being relevant to planet Earth now. It's time to update and modernize our society, and at a much faster pace!

Excluded people have been left uneducated, tricked, and lied to, and so many believe the supernatural religious claims and/or the many very unlikely popular scientific claims. But I think that the time is coming where many excluded people, are going to learn the extent of how they are being lied to, and get much better at detecting lies and determining who usually tells the truth and who usually lies. Many times those poor victims of all the lies and denial of service are the most willing to listen, while those who get D2BW generally try to avoid those who don't. Most of the D2BW consumers don't like to talk to excluded people because they already know that most of what the public is told about crimes, religions, and scientific theories are deliberate lies, and because they have already seen the thought-images and heard any thought-sounds that a person denied D2BW might want to share with them.

We can only imagine how many amazing, beautiful, and important truths and people have been ignored, or have never been seen or heard because of the popularity of the inaccurate theories of religions, psychology, fraudulent "science", because of the neuron lie, and because of the widespread belief in the myth of privacy.

Chapter 9
D2B Related Music

Many song writers are very talented people that care deeply for their fellow humans, and so naturally, they hint about injustices, such as the D2B segregation. There is simply a lot of music that relates to Neuron Reading and Writing, or as I sometimes call it "Music of the BIM" (Brain Imaging Machine).

1814: "Star Spangled Banner", Francis Scott Key, "Oh say can you see...?"- no they are not allowed to say that they can see - but that era may happen around 2300 or 2400 - when a new policy is created in which neuron consumers are allowed to admit to seeing direct-to-brain™ windows freely without punishment.

1955: "The Great Pretender", The Platters, -may hint about the pretending that the D2B consumers must do- never admitting that they see and hear thoughts or know anything about anything. This song has the iconic lyric "No one can tell", and "yes!" which may hint at groups of people who vote "yes" or "no" to going public with D2B.

1956: "Don't Be Cruel" (by Otis Blackwell, recorded by Elvis Presley) –may make a double-meaning joke with the lyric "at least please telephone". For excluded people, the obvious meaning is that the singer is asking the girl to call him using the archaic telephone device. But there may be a comical double meaning, that may relate to how the guys who control D2BW (AT&T and the other telecoms) have a monopoly on seeing and hearing thought, and so coerce many poor women into sex for D2B service, or for money, all at the drop of a thought. So it is kind of a funny lyric knowing about the D2B segregation: because pleasing the people of the telephone isn't something that young women need to be asked to do - they all already pleasure the telephone neuron guys. The only people who get to participate in the rapid-moving popup-window thought-market of free and paid-for relationships ("coatings" and "coverings") are D2B owners and consumers- excluded are like drone bees that have to "master".

1962: "Do you hear what I hear?", Noel Regney and Gloria Shayne. Clearly many excluded people don't hear what D2B consumers hear, in particular, the tremendously helpful thought-audio of many other people.

1971: Gordon Lightfoot "If you could read my mind...love, what a tale my thoughts could tell" - is that the understatement of seven centuries? There is a massive secret history of thought images from millions of people, many of which may be saved in buildings owned by the phone companies and governments of Earth. We paid for most of it, but don't get to see any of it.

1978: "I'm Every Woman", Chaka Khan. "I can read your thoughts right now, everyone from A to Z". Whitney Houston might have been galvanized by Kennedy killers and like-minded people, for her covering this song, and for her affiliation with "JFK" star Kevin Costner in the movie "The Bodyguard".

1982: "Eye in the Sky", Alan Parsons Project, "I am the eye in the sky looking at you, I can read your mind. I am the maker of rules, dealing with fools, I can cheat you blind"- can and often do!

1982: "Der Kommissar", Falco. Falco was probably remotely murdered.

1983: Stevie Nicks: If Anyone Falls in Love, "from the back of your mind"

1984: Rockwell, "I always feel like somebody's watching me, and I have no privacy"

1984: "Shake You Down", by Gregory Abbott "I'm glad you picked up on my telepathy"

1984: "Almost Paradise", by Eric Carmen, "You must have read my mind" ('you must have written to my mind' - would have been an interesting lyric too), "I'll share them all with you" (all my thought-images and thought-sounds perhaps, or those of history)

1984: "Mr. Telephone Man", New Edition, "...the situation's blowing my mind..."

1986: Paul Simon: I know what I know, "It's a thing that I do from the back of my head"

1986: "You Should Be Mine" (The Woo Woo Song), Jeffrey Osborne, "...and the heat from your mind..."

1988: "What's On Your Mind?" by Information Society, "I want to know what you're thinking, tell me what is on your mind" and "If you hide away from me" may be a double-meaning on "hide" with "Hyde".

1992: "Secret World", Peter Gabriel. "All that have gone before and left no trace...making it up in our secret world", like the many Galvanizations and the secret D2B segregation.

1993: "Christmas Through Your Eyes", Gloria Estefan, "I wanna see Christmas through your eyes..." - that is now possible though 700 year old remote neuron reading technology!

1995: "Carnival", Natalie Merchant. "Hypnotized Mesmerized by what my eyes have seen..."- Imagine the violence and abuse that neuron consumers must see in their D2B or "eyes" display. "Have I been blind?"- it must be fascinating for D2B consumers to see the very few D2B excluded who wonder if there might be a massive secret thought-seeing and hearing segregation; that they may be "excluded" or D2B "blind" (and "deaf").

1996: "Don't Speak", by No Doubt "I know what you're thinking", "don't tell me cause it hurts"- may relate to how hearing people repeat their thought-audio out loud is annoying.

1999: "Amazed", by Lonestar, "I can hear your thoughts".

2000: "I Can't Make You Love Me" Bonnie Raitt (Read, Shamblin) has the line "Turn down these voices inside my head", which helps to fight against the mistaken belief that people who talk openly about hearing direct-to-brain audio have some kind of psychiatric disorder and should be restrained and drugged.

2002: "Harder to Breathe", Maroon 5, may hint about the building up of a remote suffocation ('galvanization").

2003: "Look Through My Eyes", Phil Collins - little did many of us know that many thousands of people have been looking through our eyes at the images we see for possibly hundreds of years. Our eyes are cameras, but unlike D2B consumers, we only get to see the images our eyes capture once.

2004: "Suddenly I see", by KT Tunstall, may reflect how different life is once a person gets to see videos in front of their eyes - a life of isolation, neuron victimization (although that probably continues even after a person gets to "see"), and loneliness suddenly changes into a life full of physical and intellectual pleasure and friendships with many thousands of other people.

2004: "I Miss You" - Blink-182. "Don't waste your time on me, you're already the voice inside my head" is a clear hint about remote neuron writing of sounds. Many poor excluded people may spend a lot of time thinking about D2B consumers, in no small part due to remote neuron writing. But D2B consumers only associate with other D2B consumers, and so excluded people are better off focusing on other excluded people around them for potential friends and mates.

2005: "Speed of Sound", Coldplay, "How long before I get in?", and "if you could see it then you'd understand" may be referring to D2BW.

2006: "Telephone", by Lady Gaga. A lot of neuron code words: "I can't hear", "I have got no service, in the club", "sipping that bub", "put my coat on"

Chapter 10
D2B Related Keywords

This bizarre neuron secret two-level society has created some bizarre systems. One is the insider "keyword". There are many keywords, included humans use to try and inform the poor excluded people. Here are only a very very few of them:

Keyword	Meanings
"in"	Starting a paper with "In" or "born in mind" usually indicates that the person receives direct-to-brain windows.
"out"	A vote to make a person lose direct-to-brain read and/or write service.
"excluded"/"included"	Those who are denied D2B service/those who receive some level of D2B windows.
"get"/"get out"	An insider thinks somebody should "get" (include) an outsider- but not necessary all excluded.
"dough"	A person who shills may say this to the excluded target of their shilling, to try to explain to the victim, not to take the shill personally, and that they say the rude shills just because they want the money (often called "dough").
"tenable"	It may be that 1310 was the first year that remote or direct neuron reading and writing occurred.

"dust", "fiber"	The size of the neuron reading and writing particle devices, and also possibly that D2B devices may be attached to or disguised as dust or fibers.
"end"	The D2B consumer wants the secret to end and D2B to go public, or is describing the time when that inevitability happens.
"galvanized"	Remote muscle movement - many times implies that a person was murdered by remote muscle contraction.
"you didn't get the memo"	You are excluded
"inhuman"	The vicious and callous nature of many people who get D2BW to those who are being denied.
"day in day out"	"They are in, they are out".
"echo"	How D2B excluded echo sounds and images sent to their brain.
"cover", "coat"	D2B females get coated with sperm by D2B guys.
"do you work out?"	Does an insider female take money from an excluded guy for hand - mostly it is apparently impossible - but nonetheless the phrase exists.
"you never know"	Can mean that you will always be excluded, or that a D2B consumer is never sure if a person will be included.

"sup", "soup"	Hints that you will be included when "it's soup", or when some projects of yours (for example, losing weight, or your project exposing some truth about D2BW, etc.) are complete.
"keep in mind"	Said by a D2BW consumer who is for keeping it a secret, or is referring to how something is known only to D2B consumers.
"thought", "mind"	People writing often use these words to draw attention to some aspect of D2BW and RNRAW.
"drop"	Make a person pass-out using remote neuron writing, many times to stop remote molestation.
"hold", "lung"	Remotely hold a person's lung muscle to suffocate them.
"at this time"	AT&T is somehow involved
"that voice in your head"	A good phrase to refer to how remote neuron writing corrupts people's perspectives.
"on my mind"	May imply unpleasant neuron writing. You can ask people "what's on your thought-screen?" in a humorous light.
"lies" (used in terms of space, as in "an object lies at some location")	May mean that the author is opposed to all the lies, or that all the lies are relevant to what they are writing about.
"patience"	Might be used to indicate that

	the author is claiming to have sane views, while much of the rest of the humans around them are, in their view, mentally defective, or inferior, and perhaps should be involuntarily drugged.
"will be on hand"	"hand" relating to how excluded must serve as eunuchs - forbidden from the D2BW system of finding mates. "The excluded will be on hand..." (as will their potential spouses- tee hee hee!).
"hot"	Describes popular sex videos that many D2B consumers pay to see in their eyes. Most often the videos are probably "cold" as viewed from an excluded perspective. When a person got killed it was apparently "too hot".
"...written directly to our neurons from radio and television..."	Phrases like this can be used by D2B excluded people to try to tip off other possible excluded people without sounding overly crazy, but instead sounding whimsical and sciency.
"there's no telling"	Hint about D2BW secrecy
"you can say that again"	Hint about echoing thought

Table 10.1 Table showing only a very few D2BW keywords and their associated meanings.

Chapter 11
Shills and Rebuttals

Mostly I don't respond out loud to those who "shill" (say something rude out loud to get the money offered to them in a D2B "money window"), but of course, I can't help thinking responses in my thought-audio. Many times the "shill" is directly written to our ear or thought audio and there too, generally I think of thought-audio rebuttals. Many of these rebuttals I have collected over the years, and that I had never heard of most of them before is an indication of how imbalanced the two sides are. As always, they hear your thoughts, so it's useless to waste time repeating thought-audio out loud, but also, better to take a detached analytical view of "who is paying them?", "what is their view like?", "what are the different shades of meaning in this shill?", "what is their goal?", etc. and of course just forgetting it as quickly as possible. Many shills are personalized toward each person. Many times a person takes money to shill when the word's meaning may also have a potential positive meaning.

Some classic shills and their rebuttals (best done in thought-audio only):

Shill	Rebuttal
"rat"	➢ "rot" ➢ "do unto others…" ➢ "accessory" ➢ "I never signed a non-disclosure agreement" ➢ "murderers don't deserve the right to privacy" ➢ "You show pictures of us, so it's only fair that we show pictures of you" ➢ "Free speech"/"Free info" ➢ "liars" ➢ "hogs"

	➢ "clog" (in response to "leak")
"perv", "vert", "ped"	➢ "molester", "gnat", "remote molester", "prank caller" ➢ "frigid", "cold", "antisexual", "celibates", "nuns and monks", "I'm not a celibate", "I'm single and looking" ➢ "genocide", "excludocider", "eunuchs", "bees", "drones" ➢ "in-vert", "you're that, but worse, you're murderers." ➢ "people will probably forgive your perversion, but not your violence." ➢ "You are!" ➢ "voyeur"/"voyeur vert" ➢ "kink" (funny that people can be separated into either "pervs" or "kinks"- pervs admit that they like pleasure, kinks deny it- "you are either a perv or a kink and I don't see any horns on you") ➢ "sex isn't the problem, violence is" ➢ "don't be so hard on yourself" ➢ "that's the least of your crimes" ➢ "Stop violence- that's where we all agree" ➢ "murderers!", "brutes", "thugs", "assaulters", "molesters", "liars" ➢ Yes and 1=4, and 19 hijackers done 9/11, and Oswald killed JFK, and Jesus rose from the dead, and Moses parted the sea, and the universe is expanding, and blah, blah, blah ➢ "Just make sure there's consent" ➢ "I haven't even kissed anybody in 4 years." ➢ Write any inappropriate sexual suggestions? ➢ You must have something to hide

	➤ Who are you protecting by keeping all the video secret? ➤ Blame the beamer! ➤ Ooooo! Somebody said "butt"!
"psycho", "insane", "nut", "crazy"	➤ Thought-images of: 9/11, religious people and symbols, Jesus, Muhommed, mass bowed prayers. ➤ "Fanatic", "meaniac", "sourpuss", "dull" ➤ "it's brutal" ➤ "it's torture", "it's torturous" ➤ Witch-trialer, Gulag, Siberia, torturer, just say no!, drugger, consent, habeus corpus, yer healed! ➤ "You need to draw from pseudoscience because the facts aren't on your side" ➤ "like religions" ➤ "insanity isn't the problem, violence is" ➤ "just make sure there's consent" ➤ "don't drug'm" ➤ "violento", "violentosis", "violentophrenic", "violentomania" ➤ "it's all part of the big neuron lie" ➤ "galvanazis", "camunists"
"shutup"	➤ "I could see shutting up if I was lying" ➤ "free thought and speech forever, fascism never" ➤ "you shutup!", "after you!"
"dork", "nerd"	➤ "antiscience", "anti-education", "crude", "rude", "violenter", "drop-out", "bully", "fanatic", "zelot", "jealous", "Spartan", "thug"

"God"	➢ "yeah, God of lies!", "of mass murder", "of remote molestation", "of sell-out-your-own-mother greed"
"whoa"	➢ "yeah whoa on any slow-down of D2B going public"- "yeah whoa on the violence and remote molestation"-"yeah whoa on the anti-pleasure ferver"
"tear" itch	➢ (sarcastically) "wow that's hot"
"gay"	➢ "It's ok if yer gay!", "aggressive tough-guy gay!", "sourpuss", "thug", "brute", "meaniac", "violenter", "antigay" , "not-say!"
"trust"	➢ "Yeah I trust you to lie all the time on a dime, and to stay silent about a million violent crimes."
"How's going?"	➢ "scummin' along", "fine"
generic shill rebuttal	➢ "Shill", "greedy", "scum", "molester", "every single penny (ESP!) to the victims", "you only lose money when you shill", "low brow", "scam" ➢ Good to know that they get D2BW and what kind of money windows they have. ➢ Sticks and stones may break my bones but words will never hurt me
remote molestation rebuttal	➢ Any time you are remotely molested, think of an image of a money sign ($), because sometime in the future you may receive money for this abuse, and it

	reminds them of this. ➢ "you're only making me richer" ➢ I know I don't want to be remotely molested for the rest of my life, and I don't want my kids to be remotely molested for the rest of their life. ➢ In your mind: zap them with a laser, drop them to unconsciousness, hold their lung, remove their D2B write and read, fine them, and/or vote all money to the victims ➢ For those excluded that bit on a sex offense "I molested *once*", "You are a hundred-count molester" ➢ "10,001 wrongs don't make a right" ➢ Image of television or D2B windows being turned off. ➢ "If I got D2B I would be watching my own spouse and kids." ➢ "Oh that's right, criminals beam on our head all day and we have to constantly fire back at them with our mind." ➢ "Don't start!"

Table 11.1 Shills and possible rebuttals.

Chapter 12
Included/Excluded Differences

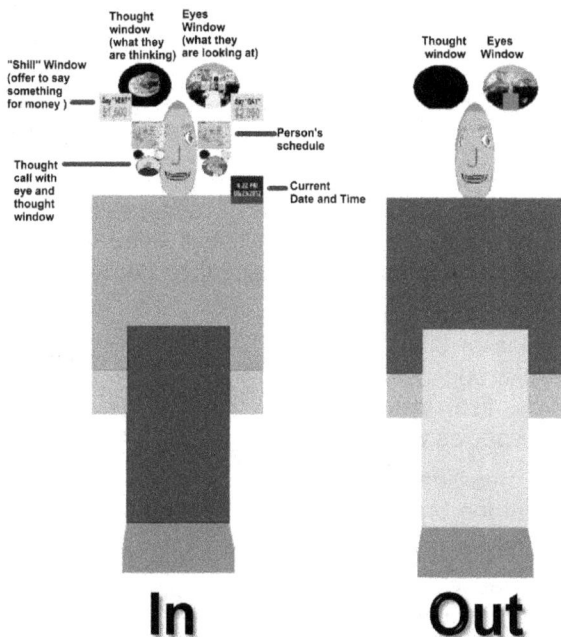

Figure 12.1 In and Out. Notice that they are both looking at a person in front of a mountain scene. The D2B consumer is thinking about a burger and fries.

There really is a caste-like, class system with typical class-based differences that has arisen because of the hording of neuron reading and writing technology by a small minority of ultra-wealthy humans, and also because of the fact that those hording this technology forbid even talking about it openly. There are serious differences between insiders and outsiders- although they are not physical as much as they are societal. Here is a table that lists important differences between those who do not receive direct-to-brain windows and those that do:

Do not receive direct-to-brain windows	Do receive and/or own direct-to-brain windows
Do not receive direct-to-brain windows (see no windows or video squares in front of their eyes)	Do receive direct-to-brain windows (can see windows and video squares in front of their eyes)
Can't voyeuristically see and hear video of people inside their homes	Can voyeuristically see and hear video of people inside their homes
Can't remotely move the muscles of brain owners: (people, dogs, cats, birds, lizards, fish, insects, and other species with brains)	Can remotely move the muscles of brain owners.
Can't hear ears (sounds of what the brains around them hear)	Can hear ears
Can't see eyes (video of what brains see)	Can see eyes
Can't hear thoughts (sounds of what brains think)	Can hear thoughts
Can't see thoughts (images of what brains think)	Can see thoughts
Can't remotely send sounds to the ears of brains	Can remotely send sounds to the ears of brains
Can't remotely send images to the eyes of brains	Can remotely send images to the eyes of brains
Can't remotely send sounds to the thoughts of brains	Can remotely send sounds to the thoughts of brains

Can't remotely send images to the thoughts of brains	Can remotely send images to the thoughts of brains
Have to masturbate	Never have to masturbate
Probably have no sexual partner	Definitely have at least one sexual partner
Probably will not reproduce	Probably will reproduce
Last to be hired, first to be fired	First to be hired, last to be fired
Earn low wages	Earn high wages
Probably are poor	Probably are wealthy
Don't get to receive dream videos while sleeping	Do get to receive dream videos while sleeping
Have their muscles remotely moved by unseen D2B consumers and/or owners	Don't have their muscles remotely moved by unseen D2B consumers and/or owners as often
Are made to itch by unseen D2B insiders using particles	Are not made to itch by unseen D2B insiders using particles as often
Are remotely sexually and nonsexually molested by unseen D2B insiders using particles	Are not remotely sexually and nonsexually molested by unseen D2B insiders using particles as often
Can't buy time on radio or television for their song or show through D2B thought-net	Can buy time on radio and television for their song or show through D2B thought-net
Can't get extra money through money-windows	Can get extra money through money-windows
Can't be published	Can be published
Can't be famous in film,	Can be famous in film,

television, music, or radio	television, music, or radio
Outsiders	Insiders
Excluded	Included
Out	In
Prey	Predators
Tutsi	Hutu
Watched	Watchers
Stalked	Stalkers
Book readers	Neuron rapers
Mostly tell the truth	Mostly tell the lie
Can talk about remote and direct neuron reading and writing	Mostly can't talk about remote or direct neuron reading or writing
Watch television and cable	Don't watch television and cable
Have to go to movie theaters to see new movies	Get new movies sent directly to their eyes wherever they are
Have to read a newspaper or watch television news to get news.	Get much more important news, and then directly to their eyes.
Have to use Google and the Internet	Don't have to use Google and the Internet
Are often sent terrible suggestions to do violence and sexually inappropriate actions	Are not often sent terrible suggestions to do violence and sexually inappropriate actions
Probably think voice in head is from God or naturally occurring, not from evil idiot neuron rape-writers	Would know that voice in head is not from God or naturally occurring, but is from evil idiot neuron rape-writers
Eunuchs, celibates,	Many partners, aroused

denied affection, veal, monks, nuns	by denying excluded affection, aroused by molesting with remote neuron writing, teasing, bullying, and tempting excluded
Empty cup, live on drops and crumbs	Cup spills over, life of decadence and excess
Drone worker bees	Queen bees

Table 12.1 Table showing differences between those who do not regularly receive D2BW and those who do receive and/or own D2BW.

And this caste system is set in stone - because the differences are so stark. It's hard for a D2B excluded person to remember this truth. There simply are too many differences between those who get direct-to-brain windows and those who are excluded for there ever to be any mixing. It seems not out of the realm of the possible that an excluded and included could have a lasting relationship, but it must be extremely rare. Mostly an included is never going to choose to even befriend an excluded- the analogy is like a Jewish human and non-Jewish human in Nazi Germany, and a black person and a white person in a slave state during slavery - there might be mixing, but it mostly doesn't happen. So as an excluded, you have an uphill battle of trying to find another person to love and make a family with, in particular when all the time, you are receiving negative and misleading remotely written feelings, and thought-audio from included people who (if they exist at all) you probably never will see, and who are definitely already in relationships with other included people, while remote neuron writing is applied to you and those you interact with to stop any potential friendships and reproduction in order to make the excluded extinct.

Chapter 13
Overvalued/Undervalued

One product of the neuron lie is that there are many people that are extremely overvalued by the excluded public, and other people that are extremely undervalued.

Overvalued	Undervalued
Religious leaders (pope, cardinals, televangelists, preachers, priests, reverends, nuns)	Scientists, teachers, and atheists
People who believe in the claims of Psychology.	Critics of Psychology
People who support the theory of relativity and expanding universe theory.	Critics of the theory of relativity and expanding universe theory.
People who say that Osama Bin Laden and 19 hijackers brought down the 3 World Trade Center buildings.	People who say that the 3 World Trade Center buildings were destroyed by controlled demolition.
People who claim that Oswald killed JFK from the School Book Depository building.	People who claim that a person in a police uniform (Frank Fiorini) killed JFK from the front while standing behind the fence.
People who claim that Sirhan Sirhan killed RFK.	People who claim that Thane Cesar killed RFK.
People who say nothing about remote neuron reading and writing or direct-to-brain windows.	People who talk openly about remote neuron reading and writing and direct-to-brain windows.

Supporters of involuntary health care treatments.	Those who oppose involuntary health care treatments.
Anti-pleasure, anti-nude, anti-sexual, anti-pornography, anti-homo- and bi-sexuality, anti-legal prostitution	Pro-pleasure, pro-nude, pro-sexual, pro-pornography, pro-homo- and bi-sexuality, pro-legal prostitution
People that believe in the supernatural claims of religions, superstitions and in the existence of non-material objects.	People that openly reject the supernatural claims of religions, superstitions, and in the existence of non-material objects.
People who want to restrict the free flow of information with copyright, patent, trademark, secrecy, privacy.	People who are opposed to copyright, patent, trademarks, secrecy, privacy, and any punishments for information crimes.
People for jailing drug buyers, sellers and users for a long time.	People against jailing drug buyers, sellers and users for a long time.
People against letting the public get to vote directly	People who support the idea of letting the public vote directly

Table 13.1 Table showing overvalued and undervalued people.

Chapter 14
Benefits of Going Public with RNRAW

There are many basic benefits of making microscopic camera, microphone, communication, and remote neuron reading and writing technology public. The many murders like 9/11, 7/7, the Kennedy, Safeway, school violence like Virginia Tech and Columbine, the remote galvanizations, all are proof that the sooner we tell the public about remote neuron reading and writing and direct-to-brain windows, the more lives we will be saving. If we tell people about RNRAW and D2BW **today**, we might be saving the lives of a few million innocent people, and could stop the growing virus of violence that may ultimately end life on Earth.

As a person who is being denied D2BW I can't possibly know them all, but here are some obvious benefits:

1. Pain a body feels might be stopped simply by a neuron writing device.
2. Blind people might have a possibility of seeing if a hovering micro-camera can write an image to their neurons.
3. Deaf people might be able to hear if sounds from a hovering microphone can be written to their neurons.
4. People might be resuscitated by neuron writing by contracting the muscles controlling their lungs and heart, and kept alive at great cost savings compared to current public methods.
5. Many murderers on the loose might be identified and punished for their violent crimes if many people could see the recorded images from the many microscopic cameras around Earth, and could see and hear thoughts. In particular, many remote particle murders, assaulters, and molesters (some with perhaps thousands of counts), could possibly be identified, stopped, and punished.
6. Many great scientific advances currently kept secret might be learned and publically utilized, like alternative electricity, combustion (light particle

releasing), and bulk atomic transmutation processes.

7. Many sexually frustrated people might be able to more easily find mates, converting that frustration away from violence.

8. Would end a lot of "remote neuron puppetry". Many excluded people might no longer be victim to the "voice of God" being neuron written to their head by the current criminal controllers of neuron writing, which is constantly used to misdirect their lives. This benefits the public tremendously, because the poor D2B excluded people will not be so easily used as remote control neuron writing puppets to do crimes.

9. Obese people could be made to not feel hunger and would probably lose weight; malnourished people could be made to feel hunger. Possibly nano-devices could even simply remove fat.

10. The speed of communication would be greatly increased, by seeing and hearing thought; great societal progress could be more rapidly made. Watching movies, searching the web, and video calls could be done while exercising without needing a computer or phone.

11. Many lies and money scams would be revealed by everybody else being able to see and hear thoughts. The majority of people would be protected from dishonest people and money scams. Those who would commit such crimes would not commit crimes knowing that many people can see them.

12. Violent people might be stopped and held in the act of violence by nearby neuron writing devices. This would stop many assaults that with the neuron writing a secret, although those neuron devices are definitely already there, normally result in murder. Modern crime solvers in governments all know who the murderers are through D2BW, the challenge for them currently is how to assemble non-D2BW evidence of the violent crimes that can be made public in court. Isn't that a wonderful, healthy and logical system? Thank goodness for "privacy" (sarcasm).

13. Remote neuron writing can make a person feel sexually aroused more easily.

14. Many people would know instantly if a person is not interested in having a relationship with them and would not waste time pursuing them. Many people who really are lonely and looking for a mate would not be lonely anymore, and would more easily find a good match.

15. Many people who have lied and their lies would be exposed to all.

16. Movies would be more entertaining - seeing a movie fully immersed - shown on every pixel in your eyes is more like being in the movie.

17. Movies can entertain and educate people with movies sent directly to their eyes and ears while they sleep - stops the monstrous criminal idiots who send nightmarish, torture dreams (like with violence, weapons, insects, on high cliffs, etc.).

18. The phone company and other neuron owners might increase their wealth by selling direct-to-brain services to more people.

19. Ends the spread of communicable disease and the deliberate infecting of excluded political enemies. Instead of neuron consumers simply watching, and neuron writers directing unwanted infections, humans could quickly determine if a person has a communicable disease.

20. Ends the program of extermination of excluded people who were excluded for telling the truth, because of their beliefs, etc. They would then have a better chance of finding dates, sex, and reproducing. Over the course of many centuries more honest people might be born on Earth.

21. Ends a lot of coercion of D2B consumers addicted to and dependent on seeing and hearing thought. Presumably the market would open for receiving D2BW and people would not have to mindlessly obey the owners of a single D2BW company.

22. Nanodevices that may already be able to slow and/or stop cancer cell growth could be used

publicly. Nanodevices that may even be able to slow or stop aging could be made public. It may be very simple to stop or even reverse the development cycle in a multicellular organism. Isn't this a good reason to stop all funding of religions and pseudosciences and focus a lot of funding of making public and developing nano-scale devices that can end aging? Perhaps all it takes is a few little changes to nucleotide order, to change the order of proteins that are activated by introns. In fact, one of the most far out ideas in this book is that, there may already be, or may be sometime soon in the near future, some group of people, presumably D2B owners and developers, who are already "ever-living"- that is, the DNA in all of their cells was changed so that they do not develop the later stages of aging normally encoded on the strands of DNA code. If you think this theory through, one requirement would apparently be that they would have to live in a way that humans that do age could not see or identify them, or else, people in the public would recognize that they never age. Probably they would need an endless supply of food and water though, and so, presumably, would need some kind of contact with those who age. Certainly they would have to be very wealthy from RNRAW and D2BW. Perhaps they have already moved off of the Earth and we simply never were shown. Or perhaps they deal with "aging" people who agree to keep the secret, or they use robots to deal with people who age normally. Perhaps they buy the silence of aging people by allowing them some partial age reversal. What would you do, in their place? There must be secret places many D2B people know about where advances in technology are secretly developed and leaked out from.

23. Even just making public nano and micro meter scale cameras and microphones would go a long way to helping the public recognize that privacy is a myth, and help to stop and catch a lot of violent humans still on the loose. These devices would also

make life much easier. In addition, such tiny cameras and devices might be able to be sent at very high speeds to other planets and stars. In fact, for all we know, there are nanometer devices already here around our star from life of other stars-because those are the kinds of devices we will probably send to other stars when we finally emerge from the backwards dark ages of religions, antipleasure, and antiscience ferver.

Chapter 15
Conclusions – What Can We Do?

Because those who own remote neuron reading and writing really have an immense advantage, any kind of serious change for the better seems unlikely in the near future, but just knowing about remote neuron reading and writing can go a long way to helping to protect yourself. The key areas that need to change are obvious, we need to stop violence, open up complete free information, and make full democracy where we vote directly on the laws we have to live under.

One thing we can do is to make movies showing what has happened. Seeing direct-to-brain windows rendered in 3D and what the view is like for a D2B consumer can show excluded people instantly what has happened – much faster than text can. One example is my video "Excluded man walks past included woman", but there are many more things to show. Being excluded from D2BW is a perfect opportunity to publicly show and tell all the important truths that those who receive D2BW all know, but are too afraid to show and tell.

One wonderful feeling about making and trying to distribute this book is that with each book sent, some poor victim of these monstrous lies and remote neuron writing is going to instantly learn many of these great truths.

So much of this work, that I have spent many long, lonely hours, by myself, compiling, under constant remote molestation, seems so unnecessary. If only those in power could have told the public all of this long ago. How terrible to be in this situation, what a terrible approach, what a cowardly and idiotic path for those powerful and wealthy people of the past, present, and future to be taking- to continue these lies, secrets, and mistaken beliefs.

I could keep writing this book forever because there is an endless amount of good and helpful information that most people excluded from direct-to-brain windows haven't been shown yet. For much more information see my web page: tedhuntington.org where I make many of my original music, videos, and writings about direct-to-brain windows and remote neuron reading and writing freely available. There are thousands of free videos of me chatting all about D2B, the segregation, and many science secrets and lies, on my video log at tedhuntington.org/vlog.

In conclusion, it's shocking to me, and I'm sure to many other people, that humans could do this to other humans- to keep direct-to-brain windows a secret for so many centuries, to not only exclude the majority of other humans on earth, but to routinely abuse them by writing to their neurons while forbidding the poor "blind and deaf" victims to even be warned that somebody might be writing to their neurons. There are so many shocking and terrible aspects. The "denial of D2B" ideological extermination: how those who are excluded can't get even a kiss, let alone sex, or to reproduce, because they are denied the right to see and hear thought; and how neuron writing is viciously used to the fullest extent to destroy any kind of relationship among excluded people. I find particularly stomach turning, the false and fraudulent theories disguised as science, like the expanding universe, time dilation, that light is not a material particle and the basis of all matter, etc. – all lies that many educated direct-to-brain owners and consumers know are not true- being proudly and widely proclaimed by the vast majority. But we can put this shocking segregation, lying, and tricking of the innocent into perspective when we realize that many humans have not only been excluded from even knowing about or receiving direct-to-brain windows, and many basic science truths, but have been simply murdered without any kind of punishment happening to the murderers. Nine-Eleven, the recent Norway and Safeway shootings, the Kennedy murders, the many black people killed under slavery, the many

Jewish, homosexual, "mental", and democratic people murdered under the Nazis, the brilliant scientists and other innocent people murdered in the Inquisition and witch trials – and of course, there must be many millions, certainly a holocaust of immense proportion, of humans who have been murdered secretly by particle beams. Nonetheless, it still shocks the mind, to experience the unbelievable nature of the two-sided society of those who get direct-to-brain windows and those who are being denied this basic service.

It seems likely that the centuries of direct-to-brain owners and consumers getting away with murder will probably come to an end some time in the future, because of the growth of cameras and free flow of information. Eventually big murders and lies like 9/11, the killing of the Kennedy's, etc. will probably not go unsolved or wrongly solved for decades, or even happen, because so many people will see the plan on the would-be murderer's thought-screen long before any violence could occur. Walking robots and cameras will soon fill the streets, and the idea of the majority of people not being able to see something you do or even images and sounds you think, will become less and less likely.   Currently many millions, even many poor people, already see and hear thought as D2B consumers, and it seems very likely that, like the number of people using the Internet, and the speed of the Internet, these numbers will only increase in time.

www.ingramcontent.com/pod-product-compliance
Lightning Source LLC
Chambersburg PA
CBHW060004210326
41520CB00009B/814